SCHAUM'S OUTLINE OF

THEORY AND PROBLEMS

OF

ELECTRONIC COMMUNICATION

Second Edition

·

LLOYD TEMES, Ph.D., P.E.
Department of Electric Technology
College of Staten Island
City University of New York

MITCHEL E. SCHULTZ
Electronics Instructor
Western Wisconsin Technical College

·

SCHAUM'S OUTLINE SERIES

McGRAW-HILL

*New York   St. Louis   San Francisco   Auckland   Bogotá   Caracas
Lisbon   London   Madrid   Mexico City   Milan   Montreal   New Delhi
San Juan   Singapore   Sydney   Tokyo   Toronto*

LLOYD TEMES has a Ph.D. and a P.E. He is at the Department of Electric Technology at the College of Staten Island, New York.

MITCHEL E. SCHULTZ, author of *Problems in Basic Electronics*, *Electric Circuits: A Text and Software Problems Manual*, and *Electronic Devices: A Text and Software Problems Manual*, earned an Associate Degree in Electronics Technology at Winona Technical College in Winona, Minnesota. He has several years of experience as an electronic technician and consultant in the RF communications field. Mitchel has taught electronics technology for the past 17 years and is currently teaching electronics at Western Wisconsin Technical College in La Crosse, Wisconsin. He also is an avid amateur radio operator and has his Amateur Extra Class License.

Schaum's Outline of Theory and Problems of
ELECTRONIC COMMUNICATION

1 2 3 4 5 6 7 8 9 10 11 12 13 14 15 16 17 18 19 20 PRS PRS 9 0 2 1 0 9 8 7

ISBN 0-07-063496-3

Sponsoring Editor: Barbara Gilson
Production Supervisor: Pamela Pelton
Editing Supervisor: Maureen B. Walker

Library of Congress Cataloging-in-Publication Data

Temes, Lloyd.
    Schaum's outline of theory and problems of electronic
communication / Lloyd Temes, Mitchel E. Schultz.—2nd ed.
      p. cm.—(Schaum outline series)
    Includes index.
    ISBN 0-07-063496-3 (papcr)
     1. Telecommunication.   2. Electronics.   I. Schultz, Mitchel E.
  II. Title.
  TK5101.T37   1998
  621.382′076—dc21   97-40617
                               CIP

McGraw-Hill
A Division of The McGraw·Hill Companies

Preface

This book is intended to be used as a supplement to any textbook covering Electronic Communications. The prerequisites for using this book include basic courses in DC and AC theory as well as a complete course covering the theory and operation of electronic devices. This book will prove extremely helpful for students enrolled in an Electronic Communications course in a two-year technical college, and it should also be quite helpful for students in a four-year electronic engineering program.

The book is divided into six chapters which are: Characteristics of Tuned LC Circuits; RF Oscillators, PLLs, and Frequency Synthesizers; Amplitude Modulation; Frequency Modulation; Transmission Lines; and Antennas. It is assumed that the student will use an electronic communications textbook for in-depth discussions of each of the major topics covered. The main purpose of this book is to develop effective problem-solving skills and to help the student study quickly and effectively. The book includes several examples and solved problems which provide the complete worked-out solution to a given problem. At the end of each chapter, there are also several supplementary problems for students to work out on their own.

LLOYD TEMES
MITCHEL E. SCHULTZ

Contents

Chapter 1

Characteristics of Tuned LC Circuits

INTRODUCTION

For any series or parallel LC circuit, the inductive reactance X_L and capacitive reactance X_C will be equal at some frequency. The frequency at which $X_L = X_C$ is called the resonant frequency. When the values of L and C are known, the resonant frequency can be calculated as:

$$f_0 = \frac{1}{2\pi\sqrt{LC}}$$

(1.1)

where f_0 represents the resonant frequency.

In general, large values of L and C provide a relatively low resonant frequency, whereas smaller values of L and C provide a higher resonant frequency. The most common application of resonance is in radio-frequency (RF) circuits where tuning is important. Tuning refers to an LC circuit's ability to provide maximum voltage output at the resonant frequency compared with the voltage output at frequencies either above or below resonance. More specifically, tuning is used in RF circuits when it is desired to pass only a specific band or channel of frequencies while at the same time completely rejecting or blocking all others. The use of tuned LC circuits is found in every television, video cassette recorder (VCR), AM/FM receiver, and satellite to name just a few of the more popular applications.

1.1 CHARACTERISTICS OF A SERIES RESONANT LC CIRCUIT

Figure 1-1 shows a series LC circuit. The series resistance r_s is a representation of the coil's own internal resistance. Because the circuit does contain some series resistance, it can be considered a series RLC circuit. Since the values of L and C are given as 25 μH and 162.1 pF respectively, the resonant frequency f_0 can be calculated as follows:

$$f_0 = \frac{1}{2\pi\sqrt{LC}}$$

$$= \frac{1}{2 \times \pi \times \sqrt{25\,\mu\text{H} \times 162.1\,\text{pF}}}$$

$$= 2.5\,\text{MHz}$$

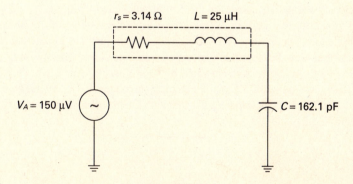

Fig. 1-1

At the resonant frequency f_0 of 2.5 MHz, the values of X_L and X_C are equal. To prove this, let us calculate each value separately. Since $X_L = 2\pi fL$ and $X_C = 1/2\pi fC$, the calculations are:

$$X_L = 2\pi f_0 L$$
$$= 2 \times \pi \times 2.5\,\text{MHz} \times 25\,\mu\text{H}$$
$$= 392.7\,\Omega$$
$$X_C = \frac{1}{2\pi f_0 C}$$
$$= \frac{1}{2 \times \pi \times 2.5\,\text{MHz} \times 162.1\,\text{pF}}$$
$$= 392.7\,\Omega$$

Because X_L and X_C are equal in magnitude and 180° out of phase, the net reactance X is 0 Ω at the resonant frequency f_0 of 2.5 MHz. Therefore, the only factor which limits the current in the circuit is the series resistance r_s of the coil. With just the low series resistance of the coil limiting the current flow, the generator voltage V_A produces the largest amount of current in the series RLC circuit at the resonant frequency. Since $r_s = 3.14\,\Omega$, the series current I at resonance can be calculated as:

$$I = \frac{V_A}{r_s}$$
$$= \frac{150\,\mu\text{V}}{3.14\,\Omega}$$
$$= 47.8\,\mu\text{A}$$

Above or below the resonant frequency of 2.5 MHz, the series current I decreases from its maximum value of 47.8 μA at resonance. The reason is that, when the generator frequency is above or below the resonant frequency of 2.5 MHz, the net reactance X is no longer zero and Z_T increases. Above the resonant frequency, $X_L > X_C$ and the net reactance X is inductive. Below the resonant frequency, $X_C > X_L$ and the net reactance X is capacitive. To calculate the total impedance Z_T of an RLC circuit at any frequency, use the following equation:

$$\boxed{Z_T = \sqrt{R^2 + X^2}} \tag{1.2}$$

where X represents the net reactance of the circuit calculated as either $X_L - X_C$ or $X_C - X_L$ depending whether X_L is greater or less than X_C. Here R represents the total series resistance in the circuit.

Keep in mind that, for any series RLC circuit, Z_T is minimum and equal to the value of R at the resonant frequency f_0. Furthermore, because the total impedance Z_T is purely resistive at f_0, the phase angle between the generator voltage and series current must be 0° at this frequency. Above f_0, the series RLC circuit appears inductive, and the current I lags the generator voltage V_A. Conversely, below f_0, the series RLC circuit appears capacitive and I leads V_A. To calculate the phase angle between V_A and I at any frequency, use the following equation:

$$\boxed{\Theta_Z = \arctan \frac{X}{R}} \tag{1.3}$$

where the subscript Z indicates that Θ_Z is found from the impedance triangle of a series RL or RC circuit.

Series Resonant Response Curves

Figure 1-2 shows a general graph of Z_T and I versus frequency for any series RLC circuit. Notice the graph of I versus f is represented as a solid line, whereas the curve corresponding to Z_T versus f is represented as a dashed line. As you can see from the graph of I versus f, the series current I is small below f_0 and then increases to its maximum value at f_0. Above f_0, the series current I again decreases to a small value. The opposite is true for the dashed curve representing Z_T versus f. In this case, Z_T has a high value below resonance with its value decreasing as the resonant frequency is approached. At f_0, Z_T has its minimum value equal to the circuit's series resistance. Above f_0, Z_T increases to a much higher value. Keep in mind that Z_T is larger above and below f_0, since X_L and X_C are not exactly equal and therefore do not completely cancel.

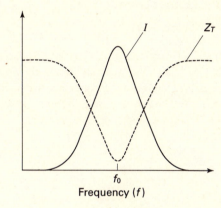

Fig. 1-2

Q of a Series Resonant Circuit

The quality or figure of merit of a series resonant circuit is indicated by a factor known as Q. In general, the larger the ratio of reactance to resistance at resonance, the higher is the circuit Q and the more pronounced is the resonant effect. For a series RLC circuit in which the coil resistance r_s is the only series resistance, the circuit Q can be calculated as:

$$Q = \frac{X_L}{r_s}$$

(1.4)

It is important to note that Q is a numerical value without any units, because it is a ratio of reactance to resistance in which the ohms unit cancels. Also, Q can be calculated using X_C instead of X_L, since both have the same value at resonance. However, the circuit Q is usually considered in terms of X_L, since the coil often contains the only series resistance in the circuit. In general, a circuit Q of 10 or more is considered high, whereas a circuit Q less than 10 is considered low. As you can see from Eq. (1.4), the value of Q can be decreased by increasing the amount of series resistance. Since series resistance cannot practically be removed from the original circuit, however, the only way to increase Q is to somehow increase the value of X_L at f_0. One way to do this is to use as high an L/C ratio as is practical. For example, if (in a given circuit) L is doubled and C is halved, then f_0 does not change but X_L and X_C each double in value. Assuming the series resistance remains the same, Q doubles.

It may appear from the equation for Q that its value can increase without limit as X_L increases for higher frequencies. This is not the case however, since such effects as skin effect, eddy currents, and hysteresis losses produce some increase in the coil's resistance at higher frequencies.

Q Rise in Voltage at Resonance

At the resonant frequency, the inductor and capacitor voltages are maximum. This is because the series current I is also at its maximum value at resonance. To calculate the inductor voltage V_L and the capacitor voltage V_C at any frequency, simply multiply the series current I by the values of X_L and X_C respectively. This can be shown as:

$$V_L = I \times X_L \qquad (1.5)$$
$$V_C = I \times X_C \qquad (1.6)$$

At resonance, where $Z_T = r_s$, I is calculated as:

$$I = \frac{V_A}{r_s}$$

If V_A/r_s is substituted for I in the equation for either V_L or V_C, the equation looks like this:

$$V_L = V_C = \frac{V_A}{r_s} \times X_L$$

Rearranging V_A and X_L gives us:

$$V_L = V_C = \frac{X_L}{r_s} \times V_A$$

Since $Q = X_L/r_s$, V_L and V_C can be calculated as:

$$V_L = V_C = Q \times V_A \qquad (1.7)$$

This equation can only be used at the resonant frequency.

As you can see, the voltage across either L or C appears Q times greater than the generator voltage V_A at the resonant frequency. This effect is very important, since the output is usually taken across either L or C in a series LC circuit.

Bandwidth of a Series Resonant Circuit

Any series resonant circuit has an associated band of frequencies that produce the desired resonant effect. The width of the resonant band is determined by the circuit Q. By definition, the bandwidth BW of a resonant circuit is defined as the gap between those frequencies for which the resonant effect is 70.7% or more of its maximum value at resonance. To calculate the bandwidth of a resonant circuit, use the following equation:

$$\mathrm{BW} = \frac{f_0}{Q} \qquad (1.8)$$

As you can see, BW is inversely proportional to the circuit's Q; that is, as Q increases, BW decreases and vice versa.

The bandwidth BW of a series resonant circuit is illustrated in Figure 1-3(a). The edge frequencies, identified as f_1 and f_2, indicate the lower and upper frequencies at which the series current I has been reduced to 70.7% of its maximum value at resonance. The edge frequencies are calculated as:

$$f_1 = f_0 - \frac{BW}{2} \qquad (1.9)$$

$$f_2 = f_0 + \frac{BW}{2} \qquad (1.10)$$

To increase the bandwidth, which is sometimes desirable, the circuit Q can be reduced by adding additional series resistance to the circuit. Since resistance cannot practically be removed from the original circuit, however, the only way to increase the Q and thereby decrease the bandwidth is to increase the L/C ratio. Figure 1-3(b) compares the response curves for a resonant LC circuit with low, medium, and high Q. Notice how the bandwidth increases as Q decreases.

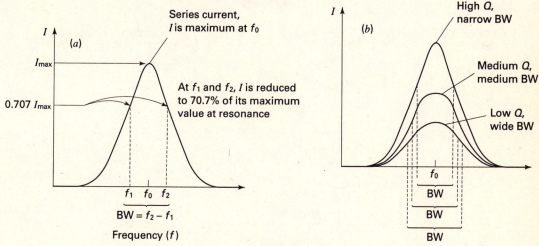

Fig. 1-3

EXAMPLE 1.1. In Fig. 1-1, calculate the following unknown quantities: Q, V_L, V_C, BW, f_1, and f_2. Assume that the generator frequency is equal to the resonant frequency f_0 of the LC circuit.

Ans. Recall that $X_L = 392.7\ \Omega$ at f_0 and that $r_s = 3.14\ \Omega$. Using these values, Q is calculated as:

$$Q = \frac{X_L}{r_s}$$

$$= \frac{392.7\ \Omega}{3.14\ \Omega}$$

$$= 125$$

Knowing Q allows us to calculate both V_L and V_C at f_0.

$$V_L = V_C = Q \times V_A$$

$$= 125 \times 150\ \mu V$$

$$= 18.75\ mV$$

To calculate the bandwidth and edge frequencies, we proceed as shown.

$$BW = \frac{f_0}{Q}$$

$$= \frac{2.5\ MHz}{125}$$

$$= 20\ kHz$$

$$f_1 = f_0 - \frac{BW}{2}$$

$$= 2.5\,\text{MHz} - \frac{20\,\text{kHz}}{2}$$

$$= 2.5\,\text{MHz} - 10\,\text{kHz}$$

$$= 2.49\,\text{MHz}$$

$$f_2 = f_0 + \frac{BW}{2}$$

$$= 2.5\,\text{MHz} + \frac{20\,\text{kHz}}{2}$$

$$= 2.5\,\text{MHz} + 10\,\text{kHz}$$

$$= 2.51\,\text{MHz}$$

EXAMPLE 1.2   Refer to Fig. 1-4. Solve for f_0, BW, f_1, and f_2. Also, solve for the following at f_0: X_L, X_C, Z_T, I, Q, V_L, and V_C.

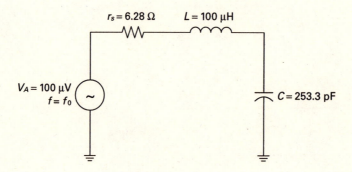

Fig. 1-4

Ans.   Begin by calculating the resonant frequency f_0.

$$f_0 = \frac{1}{2\pi\sqrt{LC}}$$

$$= \frac{1}{2 \times \pi \times \sqrt{100\,\mu\text{H} \times 253.3\,\text{pF}}}$$

$$= 1\,\text{MHz}$$

Next, calculate X_L and X_C at f_0.

$$X_L = 2\pi f_0 L$$

$$= 2 \times \pi \times 1\,\text{MHz} \times 100\,\mu\text{H}$$

$$= 628.3\,\Omega$$

$$X_C = \frac{1}{2\pi f_0 C}$$

$$= \frac{1}{2 \times \pi \times 1\,\text{MHz} \times 253.3\,\text{pF}}$$

$$= 628.3\,\Omega$$

Since $X_L = X_C$ at f_0, $Z_T = r_s$ which is 6.28 Ω in this case. Next, calculate I at f_0.

$$I = \frac{V_A}{r_s}$$

$$= \frac{100\,\mu\text{V}}{6.28\,\Omega}$$

$$= 15.9\,\mu\text{A}$$

To solve for the remaining unknowns, proceed as shown.

$$Q = \frac{X_L}{r_s}$$

$$= \frac{628.3\,\Omega}{6.28\,\Omega}$$

$$= 100$$

$$V_L = V_C = Q \times V_A$$

$$= 100 \times 100\,\mu\text{V}$$

$$= 10\,\text{mV}$$

$$\text{BW} = \frac{f_0}{Q}$$

$$= \frac{1\,\text{MHz}}{100}$$

$$= 10\,\text{kHz}$$

$$f_1 = f_0 - \frac{\text{BW}}{2}$$

$$= 1\,\text{MHz} - \frac{10\,\text{kHz}}{2}$$

$$= 995\,\text{kHz}$$

$$f_2 = f_0 + \frac{\text{BW}}{2}$$

$$= 1\,\text{MHz} + \frac{20\,\text{kHz}}{2}$$

$$= 1.005\,\text{MHz}$$

Tuning an LC Circuit

For any variable capacitor, the tuning range TR is the ratio of its maximum capacitance to its minimum capacitance. Expressed as an equation, we have:

$$\boxed{\text{TR} = \frac{C_{\max}}{C_{\min}}} \qquad\qquad (1.11)$$

For any tuned LC circuit in which the capacitance is varied, the following relationships exist.

$$f_{0(\max)} = \frac{1}{2\pi\sqrt{LC_{\min}}}$$

$$f_{0(\min)} = \frac{1}{2\pi\sqrt{LC_{\max}}}$$

Expressed as a ratio we have:

$$\frac{f_{0(\max)}}{f_{0(\min)}} = \frac{\dfrac{1}{2\pi\sqrt{LC_{\min}}}}{\dfrac{1}{2\pi\sqrt{LC_{\max}}}}$$

Squaring both sides gives us:

$$\left(\frac{f_{0(\max)}}{f_{0(\min)}}\right)^2 = \frac{\dfrac{1}{4\pi^2 LC_{\min}}}{\dfrac{1}{4\pi^2 LC_{\max}}}$$

Simplifying this equation to its simplest form gives us:

$$\left(\frac{f_{0(\max)}}{f_{0(\min)}}\right)^2 = \frac{C_{\max}}{C_{\min}}$$

Since $TR = C_{\max}/C_{\min}$, the following is true.

$$\boxed{TR = \left(\frac{f_{0(\max)}}{f_{0(\min)}}\right)^2} \tag{1.12}$$

EXAMPLE 1.3.   Suppose that an LC circuit is to be designed so that it is capable of tuning over a frequency range of 540 to 1600 kHz. Calculate the required tuning ratio of the variable capacitance.

   Ans.   The required tuning ratio of the variable capacitor is calculated as follows:

$$TR = \left(\frac{f_{0(\max)}}{f_{0(\min)}}\right)^2$$
$$= \left(\frac{1{,}600\,\text{kHz}}{540\,\text{kHz}}\right)^2$$
$$= \frac{8.8}{1}$$

EXAMPLE 1.4.   In Example 1.3, assume the value of L to be $239\,\mu\text{H}$. Calculate the values of $C_{\max}$ and $C_{\min}$.

   Ans.   Since $f_{0(\min)}$ occurs when $C = C_{\max}$, we have:

$$f_{0(\min)} = \frac{1}{2\pi\sqrt{LC_{\max}}}$$

To solve for $C_{\max}$, we proceed as follows:

$$f_{0(\min)}^2 = \frac{1}{4\pi^2 LC_{\max}}$$
$$C_{\max} = \frac{1}{4\pi^2 f_{0(\min)}^2 L}$$
$$= \frac{1}{4\pi^2 \times 540^2\,\text{kHz} \times 239\,\mu\text{H}}$$
$$= 363.4\,\text{pF}$$

To calculate $C_{\min}$, recall that:

$$\left(\frac{f_{0(\max)}}{f_{0(\min)}}\right)^2 = \frac{C_{\max}}{C_{\min}}$$

where

$$\left(\frac{f_{0(\max)}}{f_{0(\min)}}\right)^2 = \text{TR} = \frac{8.8}{1}$$

Solving for $C_{\min}$ gives us:

$$C_{\min} = \frac{C_{\max}}{8.8}$$

$$= \frac{363.4\,\text{pF}}{8.8}$$

$$= 41.3\,\text{pF}$$

Bandwidth Remains Constant When C is Varied

When C is varied in an LC circuit, the resonant frequency changes but the bandwidth remains constant. This is important because it is imperative that the bandwidth of an LC circuit remain the same throughout the tuning range. The reason that the bandwidth does not change when C is varied is that the Q of the circuit varies in direct proportion to the resonant frequency. That is, if the resonant frequency is doubled by reducing the value of C, the circuit Q also doubles. Conversely, if the circuit's resonant frequency is halved by increasing C, the circuit Q is also halved. Since BW $= f_0/Q$, it is obvious that there can be no change in the circuit's bandwidth as C is varied throughout its tuning range. If L rather than C is varied, however, the bandwidth would vary throughout the tuning range of L. Specifically, if the circuit's resonant frequency is doubled by reducing the value of L, the Q of the circuit is halved and the bandwidth is quadrupled. Conversely, if the circuit's resonant frequency is halved by increasing the value of L, Q doubles and the bandwidth is reduced by a factor of *4*. The condition of a varying bandwidth is highly undesirable. This is why an LC circuit is usually tuned by varying C rather than L.

1.2   CHARACTERISTICS OF A PARALLEL RESONANT LC CIRCUIT

Figure 1-5 shows a parallel LC circuit which is sometimes called a tank circuit. As with a series LC circuit, the resonant frequency f_0 is the frequency at which $X_L = X_C$. Unlike a series LC circuit however, the resonant effect for a parallel LC circuit is a sharp increase in the circuit's total impedance at f_0. The reason for this is that the inductive and capacitive branch currents are equal at the resonant frequency as a result of X_L and X_C being equal. Since the inductive current I_L and the capacitive current I_C are 180° out of phase, the net or total line current equals zero at the resonant frequency. With a total line current I_T of zero, the tank impedance Z_{tank} approaches infinity at the resonant frequency.

Practical LC Tank Circuit

In a practical LC tank circuit, the inductive branch impedance is slightly greater than the impedance of the capacitive branch at f_0. This is owing to the fact that a coil always contains a certain amount of internal resistance. When the coil has a Q of 10 or more, the branch currents are practically equal since the coil resistance r_s contributes very little to the overall impedance of the inductive branch. Nevertheless, I_L is always slightly less than I_C at f_0. This means that, at f_0, the net line current I_T is never exactly zero, and as a result the tank impedance is never actually infinity.

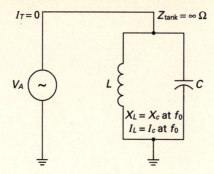

Fig. 1-5

Parallel Resonant Response Curve

Figure 1-6 shows a general graph of Z_{tank} and I_T versus f for any parallel LC circuit. Notice that the graph of I_T versus f is represented as a solid line whereas the curve representing Z_{tank} versus f is shown as a dashed line. As you can see from the curves, Z_{tank} is maximum and I_T is minimum at the resonant frequency f_0. Below f_0, I_L is greater than I_C, and the net line current I_T increases above its minimum value at f_0. Similarly, above f_0, I_C is greater than I_L, and the net line current I_T again increases above its minimum value at f_0. The fact that I_T is greater than its minimum value above and below f_0 is the reason why Z_{tank} is less than its maximum value above and below f_0.

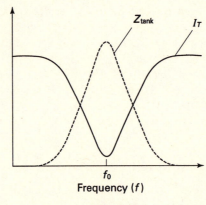

Fig. 1-6

At the resonant frequency, the tank circuit draws in phase current from the generator, which means that the phase angle of the circuit is $0°$ at f_0. Below f_0, the net line current I_T lags the generator voltage V_A, since $I_L > I_C$. Above f_0, I_T leads V_A, since $I_C > I_L$. The lagging phase angle of I_T below f_0 indicates that the tank appears inductive on the low side of resonance. Conversely, the leading phase angle of I_T above f_0 indicates that the tank appears capacitive on the high side of resonance.

Calculating the Tank Impedance at the Resonant Frequency

Refer to Fig. 1-7. To derive an equation for Z_{tank} at f_0, state each branch impedance in both rectangular and polar form. Next, use the equation

$$Z_{\text{tank}} = \frac{Z_1 Z_2}{Z_1 + Z_2}$$

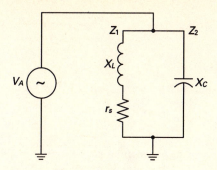

Fig. 1-7

In Fig. 1-7, Z_1 represents the impedance of the inductive branch and Z_2 represents the impedance of the capacitive branch. In the analysis that follows, assume $X_L = X_C$ since the derivation is based on the value of Z_{tank} at f_0.

If $Q \geq 10$, which is usually the case, then the following approximations can be made.

$$Z_1 = r_s + jX_L = X_L \angle{+90°}$$
$$Z_2 = 0 - jX_C = X_C \angle{-90°}$$

and

$$Z_1 + Z_2 = r_s + j0 = r_s \angle{0°}$$

Substituting the polar form of Z_1, Z_2, and $Z_1 + Z_2$ into the equation for Z_{tank} gives us the following:

$$Z_{tank} = \frac{Z_1 Z_2}{Z_1 + Z_2}$$
$$= \frac{X_L \angle{90°} \times X_C \angle{-90°}}{r_s \angle{0°}}$$

Since $Q = X_L/r_s$, this equation can be reduced to:

$$Z_{tank} = Q \times X_C$$

Since $X_C = X_L$ at f_0, Z_{tank} is usually stated as:

$$\boxed{Z_{tank} = Q \times X_L} \tag{1.13}$$

With Z_{tank} known, the net line current I_T can be calculated as:

$$\boxed{I_T = \frac{V_A}{Z_{tank}}} \tag{1.14}$$

Expanding the denominator in Eq. (*1.14*) allows us to state a relationship between the net line current I_T and the individual branch currents I_L and I_C at resonance.

$$I_T = \frac{V_A}{Q \times X_L}$$
$$= \frac{V_A}{X_L} \times \frac{1}{Q}$$

Since $V_A/X_L = I_L$, we have:

$$I_T = \frac{I_L}{Q}$$

Since $X_C = X_L$ at f_0, the following is also true for the capacitive current I_C.

$$I_T = \frac{I_C}{Q}$$

These relationships indicate that at resonance, the net line current I_T is Q times smaller than either I_L or I_C.

EXAMPLE 1.5.   In Fig. 1-8, calculate the following: f_0, X_L, X_C, I_L, I_C, Q, Z_{tank}, and I_T.
    Ans.   Begin by calculating f_0.

$$f_0 = \frac{1}{2\pi\sqrt{LC}}$$

$$= \frac{1}{2 \times \pi \times \sqrt{20\,\mu\text{H} \times 79.15\,\text{pF}}}$$

$$= 4\,\text{MHz}$$

Next, calculate X_L and X_C at f_0.

$$X_L = 2\pi f_0 L$$

$$= 2 \times \pi \times 4\,\text{MHz} \times 20\,\mu\text{H}$$

$$= 502.7\,\Omega$$

At f_0, $X_C = X_L = 502.7\,\Omega$. Next, calculate I_L and I_C.

$$I_L = \frac{V_A}{X_L}$$

$$= \frac{150\,\text{mV}}{502.7\,\Omega}$$

$$= 298.4\,\mu\text{A}$$

At f_0, $I_C = I_L = 298.4\,\mu\text{A}$. Next, calculate Q, Z_{tank}, and I_T.

$$Q = \frac{X_L}{r_s}$$

$$= \frac{502.7\,\Omega}{6.28\,\Omega}$$

$$= 80$$

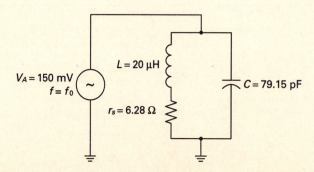

Fig. 1-8

$$Z_{\text{tank}} = Q \times X_L$$
$$= 80 \times 502.7\,\Omega$$
$$= 40.2\,\text{k}\Omega$$

$$I_T = \frac{V_A}{Z_{\text{tank}}}$$
$$= \frac{150\,\text{mV}}{40.2\,\text{k}\Omega}$$
$$= 3.73\,\mu\text{A}$$

or

$$I_T = \frac{I_L \text{ or } I_C}{Q}$$
$$= \frac{298.4\,\mu\text{A}}{80}$$
$$= 3.73\,\mu\text{A}$$

Q and Bandwidth of a Parallel Resonant Circuit

Assuming that the Q of the coil is 10 or more, a tank circuit appears as a very large resistance at the resonant frequency f_0. In effect, the formula $Z_{\text{tank}} = Q \times X_L$ transforms the series resistance r_s of the coil into an equivalent parallel resistance R_P. This is shown in Fig. 1-9. Since $X_L = X_C$ (and thus $I_L = I_C$), R_P is effectively in parallel with an open circuit. Although the value of R_P equals the tank impedance at resonance, Z_{tank} is still used instead of R_P when identifying the tank impedance at f_0. This is done to avoid confusing R_P with any other external resistance which may be connected in parallel with the tank. When there is no external load connected in parallel with the tank, the Q of the circuit is determined by the Q of the coil which is calculated as X_L/r_s. However, the Q of a tank circuit can also be stated as a ratio of tank impedance Z_{tank} to inductive reactance X_L. In other words, Eq. (*1.13*) can be reconfigured as:

$$Q = \frac{Z_{\text{tank}}}{X_L} \tag{1.15}$$

Once the Q of a tank circuit has been determined, Eqs. (*1.8*) through (*1.10*) can be used to calculate the bandwidth and edge frequencies. For a tank circuit, the edge frequencies f_1 and f_2 are defined as those frequencies at which Z_{tank} has been reduced to 70.7% of its maximum value at f_0. As before, the bandwidth includes those frequencies extending from f_1 to f_2.

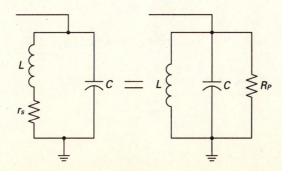

Fig. 1-9

EXAMPLE 1.6.   In Fig. 1-8, calculate BW, f_1, and f_2. Recall that $f_0 = 4$ MHz and $Q = 80$.

Ans.

$$BW = \frac{f_0}{Q}$$

$$= \frac{4\,\text{MHz}}{80}$$

$$= 50\,\text{kHz}$$

$$f_1 = f_0 - \frac{BW}{2}$$

$$= 4\,\text{MHz} - \frac{50\,\text{kHz}}{2}$$

$$= 3.975\,\text{MHz}$$

$$f_2 = f_0 + \frac{BW}{2}$$

$$= 4\,\text{MHz} + \frac{50\,\text{kHz}}{2}$$

$$= 4.025\,\text{MHz}$$

A tank circuit cannot perform a useful function all by itself. In practice, a tank is used in conjunction with other electronic circuitry to perform a useful task. The electronic circuitry to which the tank is connected serves as a load on the tank. In the discussion that follows, the load is considered to be purely resistive.

External Load Decreases Q and Increases BW

If a load resistor R_L is connected in parallel with the tank, the overall Q of the circuit becomes less than the Q of the tank itself. With a load R_L connected, more net line current I_T flows at resonance. Since the load is a resistive branch, this current cannot be cancelled by either of the reactive branch currents. The overall effect is that R_L reduces the sharpness of the resonant effect.

Since Z_{tank} is in parallel with R_L, the equivalent impedance Z_{eq} of the circuit at resonance equals $Z_{\text{tank}} \| R_L$, where "$\|$" represents Z_{tank} in parallel with R_L. When $Z_{\text{tank}} \geq 10R_L$, then $Q_{\text{ckt}} \cong R_L/X_L$, where Q_{ckt} represents the overall Q of the circuit. This is clearly stated as:

$$Q_{\text{ckt}} = \frac{R_L}{X_L} \quad (Z_{\text{tank}} \geq 10R_L) \tag{1.16}$$

When $Z_{\text{tank}} < 10R_L$, then Q_{ckt} must be calculated as:

$$Q_{\text{ckt}} = \frac{Z_{\text{tank}} \| R_L}{X_L} \quad (Z_{\text{tank}} < 10R_L) \tag{1.17}$$

With Q_{ckt} known, the bandwidth BW is calculated as before: BW $= f_0/Q_{\text{ckt}}$.

EXAMPLE 1.7.   In Fig. 1-10 (*a*), calculate f_0, X_L, Q, Z_{tank}, and BW.
   Ans.   Begin by calculating f_0.

$$f_0 = \frac{1}{2\pi\sqrt{LC}}$$

$$= \frac{1}{2 \times \pi \times \sqrt{100\,\mu\text{H} \times 63.3\,\text{pF}}}$$

$$= 2\,\text{MHz}$$

Next, calculate X_L and Q.

$$X_L = 2\pi f_0 L$$
$$= 2 \times \pi \times 2\,\text{MHz} \times 100\,\mu\text{H}$$
$$= 1.26\,\text{k}\Omega$$

$$Q = \frac{X_L}{r_s}$$
$$= \frac{1.26\,\text{k}\Omega}{12.6\,\Omega}$$
$$= 100$$

Finally, calculate Z_tank and BW.

$$Z_\text{tank} = Q \times X_L$$
$$= 100 \times 1.26\,\text{k}\Omega$$
$$= 126\,\text{k}\Omega$$

$$\text{BW} = \frac{f_0}{Q}$$
$$= \frac{2\,\text{MHz}}{100}$$
$$= 20\,\text{kHz}$$

EXAMPLE 1.8.   Figure 1-10(b) shows the same LC tank circuit as in Fig. 1-10(a). However, a 27 kΩ load has been placed across the tank. Calculate Q_ckt and BW.

Ans.   Before Q_ckt can be determined, find out whether Z_tank is 10 or more times larger than R_L.

$$\frac{Z_\text{tank}}{R_L} = \frac{126\,\text{k}\Omega}{27\,\text{k}\Omega} = \frac{4.7}{1}$$

Since $Z_\text{tank} < 10R_L$, Eq. (*1.17*) must be used to calculate Q_ckt.

$$Q_\text{ckt} = \frac{Z_\text{tank}\|R_L}{X_L}$$
$$= \frac{126\,\text{k}\Omega\|27\,\text{k}\Omega}{1.26\,\text{k}\Omega}$$
$$= 17.6$$

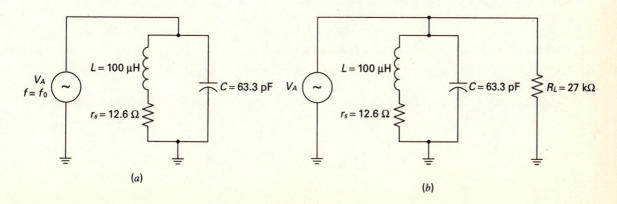

(a)   (b)

Fig. 1-10

Next, calculate the bandwidth.

$$BW = \frac{f_0}{Q_{\text{ckt}}}$$

$$= \frac{2\,\text{MHz}}{17.6}$$

$$= 113.6\,\text{kHz}$$

As you can see, the addition of a load resistor R_L increases the bandwidth. Without a load present in Fig. 1-10(a), the bandwidth was only 20 kHz. However, with the presence of a 27 kΩ load in Fig. 1-10(b), the bandwidth was broadened to 113.6 kHz.

1.3   TRANSFORMER COUPLING

A common coupling arrangement encountered in communications (RF) equipment is a coupling transformer. Figure 1-11 shows various tuned coupling arrangements in which a capacitor or capacitors are used in conjunction with a transformer to allow only the desired band of RF signals to pass. The connection of a capacitor to either the primary or secondary of the transformer forms a tuned LC circuit. Figure 1-11(a) shows a coupling transformer with a tuned primary whereas Fig. 1-11(b) shows a transformer with a tuned secondary. Figure 1-11(c) shows a double-tuned transformer in which both the primary and secondary are tuned.

The coefficient of coupling (k) of a transformer is a measure of how much of the magnetic flux originated by the primary links the secondary of the transformer. Although for audio equipment it is not unusual to find coefficients of coupling on the order of 0.90 and higher, transformers in RF service have coefficients of coupling on the order of 0.01 to 0.05. RF transformer coupling arrangements have frequency response curves as shown in Figs. 1-12 and 1-13. Refer to Fig. 1-13, the response curve for the double-tuned transformer, and note the reference to overcoupled, undercoupled, and critically coupled conditions.

The *critically coupled* situation is the one that provides maximum output and maximum bandwidth without a dip in output at the resonant frequency. For a doubled-tuned transformer, critical coupling results when the coefficient of coupling k is equal to the reciprocal of the square root of the product of the Q of the primary and the Q of the secondary:

$$k_c = \frac{1}{\sqrt{Q_P Q_S}} \tag{1.18}$$

In Eq. (1.18), Q_P represents the Q of the primary and Q_S represents the Q of the secondary.

The undercoupled situation where $k < k_c$ provides neither maximum output nor maximum bandwidth. In the overcoupled case where $k > k_c$ a dip appears in the response curve at the resonant frequency. The slightly overcoupled condition is sometimes desired because it provides steeper sides to the response curve and thus sharpens rejection of undesired signals. A coefficient of coupling k of 1.5 times the critical coefficient of coupling is frequently used and considered desirable.

The bandwidth of a double-tuned coupling transformer is equal to the product of the coefficient of coupling k and the resonant frequency f_0. This is shown as:

$$BW = kf_0 \tag{1.19}$$

EXAMPLE 1.9.   Determine the necessary value of k to provide critical coupling for a double-tuned transformer in which $Q_P = 60$ and $Q_S = 90$.

Ans.   Using Eq. (1.18), the calculations are as follows:

$$k_c = \frac{1}{\sqrt{Q_P Q_S}}$$

$$= \frac{1}{\sqrt{60 \times 90}}$$

$$= 0.0136$$

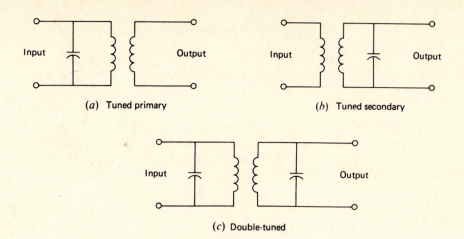

(a) Tuned primary   (b) Tuned secondary

(c) Double-tuned

Fig. 1-11

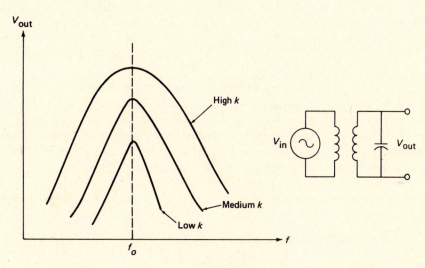

Fig. 1-12

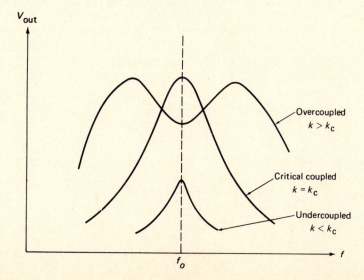

Fig. 1-13

Solved Problems

1.1 Calculate the resonant frequency f_0 in Fig. 1-14.

SOLUTION

Using Eq. (*1.1*), the calculations for f_0 are as follows:

$$f_0 = \frac{1}{2\pi\sqrt{LC}}$$

$$= \frac{1}{2 \times \pi \times \sqrt{2\,mH \times 0.001\,\mu F}}$$

$$= 112.5\,kHz$$

1.2 In Fig. 1-14, calculate X_L and X_C at f_0.

SOLUTION

Using $X_L = 2\pi f_0 L$ and $X_C = 1/2\pi f_0 C$, the calculations are:

$$X_L = 2\pi f_0 L$$

$$= 2 \times \pi \times 112.5\,kHz \times 2\,mH$$

$$= 1.41\,k\Omega$$

$$X_C = \frac{1}{2\pi f_0 C}$$

$$= \frac{1}{2 \times \pi \times 112.5\,kHz \times 0.001\,\mu F}$$

$$= 1.41\,k\Omega$$

1.3 Refer to Fig. 1-14. Calculate the total impedance Z_T at f_0.

SOLUTION

Since $X_L = X_C$ at f_0, the total impedance $Z_T = r_s$ which is 47 Ω in this case. This can be verified by using Eq. (*1.2*).

$$Z_T = \sqrt{R^2 + (X_L - X_C)^2}$$

$$= \sqrt{47^2\,\Omega + (1.41\,k\Omega - 1.41\,k\Omega)^2}$$

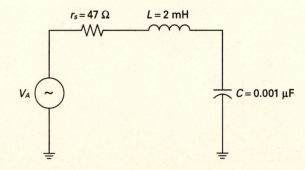

Fig. 1-14

$$= \sqrt{47^2\,\Omega + 0^2\,\Omega}$$

$$= \sqrt{47^2\,\Omega}$$

$$= 47\,\Omega$$

1.4   Refer to Fig. 1-15. Calculate the resonant frequency, f_0. Also, solve for the following at f_0; X_L, X_C, Z_T, I, V_L, V_C, and Θ_Z.

SOLUTION

Begin by calculating f_0.

$$f_0 = \frac{1}{2\pi\sqrt{LC}}$$

$$= \frac{1}{2 \times \pi \times \sqrt{400\,\mu\mathrm{H} \times 63.3\,\mathrm{pF}}}$$

$$= 1\,\mathrm{MHz}$$

Next, calculate X_L and X_C at f_0.

$$X_L = 2\pi f_0 L$$

$$= 2 \times \pi \times 1\,\mathrm{MHz} \times 400\,\mu\mathrm{H}$$

$$= 2.5\,\mathrm{k}\Omega$$

$$X_C = \frac{1}{2\pi f_0 C}$$

$$= \frac{1}{2 \times \pi \times 1\,\mathrm{MHz} \times 63.3\,\mathrm{pF}}$$

$$= 2.5\,\mathrm{k}\Omega$$

Since $X_L - X_C = 0\,\Omega$, $Z_T = r_s$ which is $25\,\Omega$ in this case.
With Z_T known, the series current I is calculated as:

$$I = \frac{V_A}{r_s}$$

$$= \frac{1\,\mathrm{V}}{25\,\Omega}$$

$$= 40\,\mathrm{mA}$$

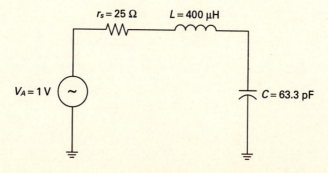

$r_s = 25\,\Omega$ $L = 400\,\mu\mathrm{H}$

$V_A = 1\,\mathrm{V}$ $C = 63.3\,\mathrm{pF}$

Fig. 1-15

To calculate V_L and V_C, use Eqs. (1.5) and (1.6).

$$V_L = I \times X_L$$
$$= 40\,\text{mA} \times 2.5\,\text{k}\Omega$$
$$= 100\,\text{V}$$
$$V_C = I \times X_C$$
$$= 40\,\text{mA} \times 2.5\,\text{k}\Omega$$
$$= 100\,\text{V}$$

To calculate Θ_Z, use Eq. (1.3).

$$\Theta_Z = \arctan\frac{X}{R}$$
$$= \arctan\frac{0\,\Omega}{25\,\Omega}$$
$$= \arctan 0$$
$$= 0°$$

1.5 In Fig. 1-15, calculate Q, BW, f_1, and f_2.

SOLUTION

To calculate Q, recall that $X_L = 2.5\,\text{k}\Omega$ at f_0. Next, use Eq. (1.4).

$$Q = \frac{X_L}{r_s}$$
$$= \frac{2.5\,\text{k}\Omega}{25\,\Omega}$$
$$= 100$$

Next, use Eq. (1.8) to calculate the bandwidth BW.

$$\text{BW} = \frac{f_0}{Q}$$
$$= \frac{1\,\text{MHz}}{100}$$
$$= 10\,\text{kHz}$$

Use Eqs. (1.9) and (1.10) to calculate f_1 and f_2 respectively.

$$f_1 = f_0 - \frac{\text{BW}}{2}$$
$$= 1\,\text{MHz} - \frac{10\,\text{kHz}}{2}$$
$$= 995\,\text{kHz}$$
$$f_2 = f_0 + \frac{\text{BW}}{2}$$
$$= 1\,\text{MHz} + \frac{10\,\text{kHz}}{2}$$
$$= 1.005\,\text{MHz}$$

1.6 Refer to Fig. 1-16. Calculate the resonant frequency f_0. Also, solve the following at f_0: X_L, X_C, Z_T, I, Q, V_L, V_C. Also solve for BW, f_1, and f_2.

SOLUTION

Begin by calculating f_0.

$$f_0 = \frac{1}{2\pi\sqrt{LC}}$$

$$= \frac{1}{2 \times \pi \times \sqrt{1\,\text{mH} \times 101.3\,\text{pF}}}$$

$$= 500\,\text{kHz}$$

Next, calculate X_L and X_C at f_0.

$$X_L = 2\pi f_0 L$$

$$= 2 \times \pi \times 500\,\text{kHz} \times 1\,\text{mH}$$

$$= 3.1\,\text{k}\Omega$$

$$X_C = \frac{1}{2\pi f_0 C}$$

$$= \frac{1}{2 \times \pi \times 500\,\text{kHz} \times 101.3\,\text{pF}}$$

$$= 3.1\,\text{k}\Omega$$

Since $X_L - X_C = 0\,\Omega$, $Z_T = r_s$ which is $62\,\Omega$ in this case. With Z_T known, the series current I can be calculated next.

$$I = \frac{V_A}{Z_T}$$

$$= \frac{250\,\text{mV}}{62\,\Omega}$$

$$= 4\,\text{mA}$$

Next, calculate Q, V_L, V_C, BW, f_1, and f_2.

$$Q = \frac{X_L}{r_s}$$

$$= \frac{3.1\,\text{k}\Omega}{62\,\Omega}$$

$$= 50$$

$$V_L = V_C = Q \times V_A$$

$$= 50 \times 250\,\text{mV}$$

$$= 12.5\,\text{V}$$

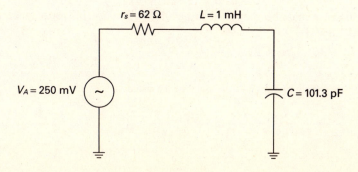

Fig. 1-16

$$BW = \frac{f_0}{Q}$$

$$= \frac{500\,kHz}{50}$$

$$= 10\,kHz$$

$$f_1 = f_0 - \frac{BW}{2}$$

$$= 500\,kHz - \frac{10\,kHz}{2}$$

$$= 495\,kHz$$

$$f_2 = f_0 + \frac{BW}{2}$$

$$= 500\,kHz + \frac{10\,kHz}{2}$$

$$= 505\,kHz$$

1.7 What size capacitance must be connected in series with a 10 μH inductance to obtain a resonant frequency f_0 of 4 MHz?

SOLUTION

Begin with the equation for f_0 and solve for the value of C.

$$f_0 = \frac{1}{2\pi\sqrt{LC}}$$

Next, square both sides of the equation.

$$f_0^2 = \frac{1}{4\pi^2 LC}$$

Next, solve for C and insert the known values of L and f_0.

$$C = \frac{1}{4\pi^2 f_0^2 L}$$

$$= \frac{1}{4 \times \pi^2 \times 4^2\,MHz \times 10\,\mu H}$$

$$= 158.3\,pF$$

1.8 What size inductance must be connected in parallel with a 250 pF capacitance to obtain a resonant frequency f_0 of 1.8 MHz.

SOLUTION

Begin with the equation for f_0 and then solve for the value of L.

$$f_0 = \frac{1}{2\pi\sqrt{LC}}$$

Next, square both sides.

$$f_0^2 = \frac{1}{4\pi^2 LC}$$

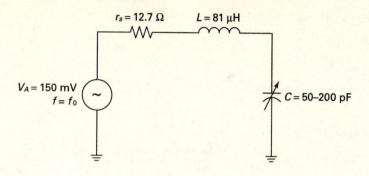

Fig. 1-17

Next, solve for L.

$$L = \frac{1}{4\pi^2 f_0^2 C}$$

$$= \frac{1}{4 \times \pi^2 \times 1.8^2 \, \text{MHz} \times 250 \, \text{pF}}$$

$$= 31.3 \, \mu\text{H}$$

1.9   Refer to Fig. 1-17. (*a*) With C set to 50 pF, calculate f_0, X_L, Q, V_C, and BW. (*b*) With C set to 200 pF, calculate f_0, X_L, Q, V_C, and BW.

SOLUTION

(*a*)   Begin by calculating f_0.

$$f_0 = \frac{1}{2\pi\sqrt{LC}}$$

$$= \frac{1}{2 \times \pi\sqrt{81 \, \mu\text{H} \times 50 \, \text{pF}}}$$

$$= 2.5 \, \text{MHz}$$

Next, calculate X_L, Q, and V_C at f_0.

$$X_L = 2\pi f_0 L$$

$$= 2 \times \pi \times 2.5 \, \text{MHz} \times 81 \, \mu\text{H}$$

$$= 1.27 \, \text{k}\Omega$$

$$Q = \frac{X_L}{r_s}$$

$$= \frac{1.27 \, \text{k}\Omega}{12.7 \, \Omega}$$

$$= 100$$

$$V_C = Q \times V_A$$

$$= 100 \times 150 \, \text{mV}$$

$$= 15 \, \text{V}$$

And finally, calculate the bandwidth.

$$BW = \frac{f_0}{Q}$$

$$= \frac{2.5\,\text{MHz}}{100}$$

$$= 25\,\text{kHz}$$

(b) First, calculate f_0.

$$f_0 = \frac{1}{2\pi\sqrt{LC}}$$

$$= \frac{1}{2\times\pi\sqrt{81\,\mu\text{H}\times 200\,\text{pF}}}$$

$$= 1.25\,\text{MHz}$$

Next, calculate X_L, Q, and V_C at f_0.

$$X_L = 2\pi f_0 L$$

$$= 2\times\pi\times 1.25\,\text{MHz}\times 81\,\mu\text{H}$$

$$= 636.2\,\Omega$$

$$Q = \frac{X_L}{r_s}$$

$$= \frac{636.2\,\Omega}{12.7\,\Omega}$$

$$= 50$$

$$V_C = Q\times V_A$$

$$= 50\times 150\,\text{mV}$$

$$= 7.5\,\text{V}$$

And finally, calculate the bandwidth.

$$BW = \frac{f_0}{Q}$$

$$= \frac{1.25\,\text{MHz}}{50}$$

$$= 25\,\text{kHz}$$

Notice how the variation in C from 50 to 200 pF did not vary the bandwidth. For both settings of C, the BW remains constant at 25 kHz.

1.10 An LC circuit is to be designed so that it is capable of tuning over a frequency range of 85 to 1145 kHz. (a) Calculate the required tuning ratio of the variable capacitance. (b) If the value of L is 120 μH, calculate the values of C_{min} and C_{max}.

SOLUTION

(a)

$$TR = \left(\frac{f_{0(max)}}{f_{0(min)}}\right)^2$$

$$= \left(\frac{1145\,\text{kHz}}{85\,\text{kHz}}\right)^2$$

$$= \frac{181.4}{1}$$

(b) To calculate the value of C_{min}, begin with:

$$f_{0(max)} = \frac{1}{2\pi\sqrt{LC_{min}}}$$

Solving for C_{min} gives us:

$$C_{min} = \frac{1}{4\pi^2 f_{0(max)}^2 L}$$

$$= \frac{1}{4 \times \pi^2 \times 1145^2\,\text{kHz} \times 120\,\mu\text{H}}$$

$$= 161\,\text{pF}$$

Next recall that, in Part (a), TR = 181.4/1. Therefore,

$$\text{TR} = \frac{C_{max}}{C_{min}}$$

$$\frac{181.4}{1} = \frac{C_{max}}{161\,\text{pF}}$$

Solving for C_{max} gives us:

$$C_{max} = 181.4 \times 161\,\text{pF}$$

$$= 29\,200\,\text{pF or 29.2 nF}$$

1.11   Refer to Fig. 1-1. Recall that $f_0 = 2.5\,\text{MHz}$, $f_1 = 2.49\,\text{MHz}$, and $f_2 = 2.51\,\text{MHz}$. Calculate Z_T, I, and Θ_Z at f_1. (All answers are carried out three places beyond the decimal point to achieve desired accuracy.)

SOLUTION

At f_1, X_L and X_C are calculated as follows:

$$X_L = 2\pi f_1 L$$

$$= 2 \times \pi \times 2.49\,\text{MHz} \times 25\,\mu\text{H}$$

$$= 391.128\,\Omega$$

$$X_C = \frac{1}{2\pi f_1 C}$$

$$= \frac{1}{2 \times \pi \times 2.49\,\text{MHz} \times 162.1\,\text{pF}}$$

$$= 394.31\,\Omega$$

At f_1, the net reactance X equals:

$$X = X_C - X_L$$

$$= 394.31\,\Omega - 391.128\,\Omega$$

$$= 3.182\,\Omega$$

To calculate Z_T at f_1, use Eq. (1.2).

$$Z_T = \sqrt{R^2 + X^2}$$

$$= \sqrt{3.14^2\,\Omega + 3.182^2\,\Omega}$$

$$= 4.47\,\Omega$$

Next, calculate the series current I.

$$I = \frac{V_A}{Z_T}$$

$$= \frac{150\ \mu\text{V}}{4.47\ \Omega}$$

$$= 33.557\ \mu\text{A}$$

Recall that the value of I at f_0 is 47.8 μA. Dividing the value of I at f_1 by the value of I at f_0 gives us:

$$\frac{I @ f_1}{I @ f_0} = \frac{33.557\ \mu\text{A}}{47.8\ \mu\text{A}} = 0.702$$

Notice that the value of I at f_1 is approximately 70% of the value of I at f_0.

And finally, the phase angle Θ_Z is calculated using Eq. (1.3).

$$\Theta_Z = \arctan\left(-\frac{X}{R}\right)$$

$$= \arctan\left(-\frac{3.182\ \Omega}{3.14\ \Omega}\right)$$

$$= -45.38°$$

Notice that $\Theta_Z = -45.38°$ at f_1, since $X = R$. Also, the phase angle is negative, since $X_C > X_L$ at f_1.

1.12 In Fig. 1-18, assume that $Q = 50$. Calculate f_0, BW, and the value of the coil resistance r_s.

SOLUTION

Begin by calculating f_0.

$$f_0 = \frac{1}{2\pi\sqrt{LC}}$$

$$= \frac{1}{2 \times \pi \times \sqrt{30\ \mu\text{H} \times 15\ \text{pF}}}$$

$$= 7.5\ \text{MHz}$$

Next, calculate BW.

$$\text{BW} = \frac{f_0}{Q}$$

$$= \frac{7.5\ \text{MHz}}{50}$$

$$= 150\ \text{kHz}$$

To calculate r_s, recall that $Q = X_L/r_s$. This formula can be arranged to solve for r_s.

$$r_s = \frac{X_L}{Q}$$

Next, solve for X_L.

$$X_L = 2\pi f_0 L$$

$$= 2 \times \pi \times 7.5\ \text{MHz} \times 30\ \mu\text{H}$$

$$= 1.41\ \text{k}\Omega$$

Solving for r_s gives us:

$$r_s = \frac{X_L}{Q}$$

$$= \frac{1.41\ \text{k}\Omega}{50}$$

$$= 28.2\ \Omega$$

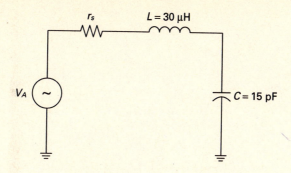

Fig. 1-18

1.13 In Fig. 1-19, calculate f_0. Also, calculate the following values at f_0: X_L, X_C, I_L, I_C, Q, Z_{tank}, and I_T.

SOLUTION

Begin by calculating f_0.

$$f_0 = \frac{1}{2\pi\sqrt{LC}}$$

$$= \frac{1}{2 \times \pi \times \sqrt{200\,\mu\text{H} \times 75\,\text{pF}}}$$

$$= 1.3\,\text{MHz}$$

Next, calculate X_L, X_C, I_L, and I_C.

$$X_L = 2\pi f_0 L$$

$$= 2 \times \pi \times 1.3\,\text{MHz} \times 200\,\mu\text{H}$$

$$= 1.63\,\text{k}\Omega$$

$$X_C = \frac{1}{2\pi f_0 C}$$

$$= \frac{1}{2 \times \pi \times 1.3\,\text{MHz} \times 75\,\text{pF}}$$

$$= 1.63\,\text{k}\Omega$$

$$I_L = \frac{V_A}{X_L}$$

$$= \frac{300\,\text{mV}}{1.63\,\text{k}\Omega}$$

$$= 184\,\mu\text{A}$$

$$I_C = \frac{V_A}{X_C}$$

$$= \frac{300\,\text{mV}}{1.63\,\text{k}\Omega}$$

$$= 184\,\mu\text{A}$$

Next, calculate Q, Z_{tank}, and I_T.

$$Q = \frac{X_L}{r_s}$$

$$= \frac{1.63\,\text{k}\Omega}{25\,\Omega}$$

$$= 65.2$$

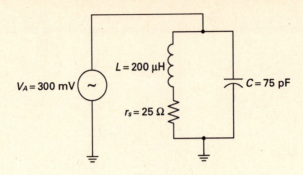

$V_A = 300 \text{ mV}$　$L = 200\ \mu H$　$C = 75 \text{ pF}$　$r_s = 25\ \Omega$

Fig. 1-19

$$Z_{tank} = Q \times X_L$$
$$= 65.2 \times 1.63 \text{ k}\Omega$$
$$= 106.3 \text{ k}\Omega$$
$$I_T = \frac{V_A}{Z_{tank}}$$
$$= \frac{300 \text{ mV}}{106.3 \text{ k}\Omega}$$
$$= 2.82\ \mu A$$
or
$$I_T = \frac{I_L \text{ or } I_C}{Q}$$
$$= \frac{184\ \mu A}{65.2}$$
$$= 2.82\ \mu A$$

1.14 In Fig. 1-19, calculate the bandwidth and edge frequencies. Recall that $f_0 = 1.3 \text{ MHz}$ and $Q = 65.2$.

SOLUTION

$$BW = \frac{f_0}{Q}$$
$$= \frac{1.3 \text{ MHz}}{65.2}$$
$$= 20 \text{ kHz}$$
$$f_1 = f_0 - \frac{BW}{2}$$
$$= 1.3 \text{ MHz} - \frac{20 \text{ kHz}}{2}$$
$$= 1.29 \text{ MHz}$$
$$f_2 = f_0 + \frac{BW}{2}$$
$$= 1.3 \text{ MHz} + \frac{20 \text{ kHz}}{2}$$
$$= 1.31 \text{ MHz}$$

1.15 In Fig. 1-19, assume that a $100\,\text{k}\Omega$ load R_L is placed in parallel with the tank. Calculate Q_{ckt} and BW.

SOLUTION

Since $Z_{\text{tank}} < 10R_L$,

$$Q_{\text{ckt}} = \frac{Z_{\text{tank}} \| R_L}{X_L}$$

$$= \frac{106.3\,\text{k}\Omega \| 100\,\text{k}\Omega}{1.63\,\text{k}\Omega}$$

$$= \frac{51.5\,\text{k}\Omega}{1.63\,\text{k}\Omega}$$

$$= 31.6$$

Next, calculate BW.

$$\text{BW} = \frac{f_0}{Q}$$

$$= \frac{1.3\,\text{MHz}}{31.6}$$

$$= 41.1\,\text{kHz}$$

1.16 In Fig. 1-20, calculate f_0. Also, solve for the following at f_0: X_L, X_C, I_L, I_C, Q, Z_{tank}, and I_T. (S_1 is open as shown.)

SOLUTION

Begin by calculating f_0.

$$f_0 = \frac{1}{2\pi\sqrt{LC}}$$

$$= \frac{1}{2 \times \pi \times \sqrt{82.7\,\mu\text{H} \times 25\,\text{pF}}}$$

$$= 3.5\,\text{MHz}$$

Next, calculate X_L, X_C, I_L, and I_C.

$$X_L = 2\pi f_0 L$$
$$= 2 \times \pi \times 3.5\,\text{MHz} \times 82.7\,\mu\text{H}$$
$$= 1.82\,\text{k}\Omega$$

$$X_C = \frac{1}{2\pi f_0 C}$$

$$= \frac{1}{2 \times \pi \times 3.5\,\text{MHz} \times 25\,\text{pF}}$$

$$= 1.82\,\text{k}\Omega$$

$$I_L = \frac{V_A}{X_L}$$

$$= \frac{10\,\text{V}}{1.82\,\text{k}\Omega}$$

$$= 5.5\,\text{mA}$$

$$I_C = \frac{V_A}{X_C}$$

$$= \frac{10\,\text{V}}{1.82\,\text{k}\Omega}$$

$$= 5.5\,\text{mA}$$

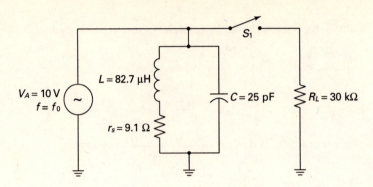

Fig. 1-20

Next, calculate Q, Z_{tank}, and I_T.

$$Q = \frac{X_L}{r_s}$$

$$= \frac{1.82\,\text{k}\Omega}{9.1\,\Omega}$$

$$= 200$$

$$Z_{\text{tank}} = Q \times X_L$$

$$= 200 \times 1.82\,\text{k}\Omega$$

$$= 364\,\text{k}\Omega$$

$$I_T = \frac{V_A}{Z_{\text{tank}}}$$

$$= \frac{10\,\text{V}}{364\,\text{k}\Omega}$$

$$= 27.5\,\mu\text{A}$$

1.17   In Fig. 1-20, calculate the bandwidth when (*a*) S_1 is open, (*b*) S_1 is closed.

SOLUTION

(*a*)   With S_1 open, the circuit Q equals the Q of the coil which was calculated to be 200 in the previous problem. Therefore:

$$\text{BW} = \frac{f_0}{Q}$$

$$= \frac{3.5\,\text{MHz}}{200}$$

$$= 17.5\,\text{kHz}$$

(*b*)   With S_2 closed, Z_{tank} is in parallel with R_L. Since $Z_{\text{tank}} > 10R_L$, the circuit Q is calculated as:

$$Q_{\text{ckt}} = \frac{R_L}{X_L}$$

$$= \frac{30\,\text{k}\Omega}{1.82\,\text{k}\Omega}$$

$$= 16.5$$

With Q_{ckt} known, the bandwidth can be calculated as:

$$\begin{aligned} \text{BW} &= \frac{f_0}{Q_{ckt}} \\ &= \frac{3.5\,\text{MHz}}{16.5} \\ &= 212.1\,\text{kHz} \end{aligned}$$

Notice how the load R_L lowers the circuit Q and broadens the bandwidth.

1.18   In Fig. 1-20, assume that S_1 is open. If the value of L is doubled and C is halved, what happens to: (*a*) f_0; (*b*) X_L, Q, and Z_{tank}; (*c*) BW.

SOLUTION

(*a*)   The value of f_0 does not change ($f_0 = 3.5\,\text{MHz}$), since the LC product remains the same.

(*b*)   Since the value of L is doubled with the same value of f_0, X_L must also double. Since the coil resistance r_s is unchanged, Q also doubles. Since $Z_{tank} = Q \times X_L$, Z_{tank} increases by a factor of 4.

(*c*)   The BW is cut in half, since Q has doubled in value.

1.19   Determine the coefficient of coupling k for a double-tuned transformer in order to provide critical coupling. $Q_P = 50$ and $Q_S = 80$.

SOLUTION

$$\begin{aligned} k_c &= \frac{1}{\sqrt{Q_P Q_S}} \\ &= \frac{1}{\sqrt{50 \times 80}} \\ &= 0.0158 \end{aligned}$$

1.20   The specifications for a double-tuned transformer to be used as a coupling network are such as to require a resonant frequency f_0 of 900 kHz and a bandwidth BW of 15 kHz. Both the primary and secondary of the transformer have an inductance of 200 μH. $Q_P = 60$ and $Q_S = 75$. Calculate: (*a*) the capacitance C required across the primary and secondary, (*b*) the required coefficient of coupling, (*c*) whether the circuit is undercoupled, critically coupled, or overcoupled.

SOLUTION

(*a*)   The required capacitance values are determined by f_0 and the inductance value of the primary and secondary coils. Since both the primary and secondary circuits require the same resonant frequency, and since in this case both primary and secondary coils have the same inductance, both circuits require the same value of capacitance C. To solve for C, transpose the equation for f_0.

$$f_0 = \frac{1}{2\pi\sqrt{LC}}$$

$$f_0^2 = \frac{1}{4\pi^2 LC}$$

$$C_1 = C_2 = \frac{1}{4\pi^2 f_0^2 L}$$

Inserting known values gives us:

$$C_1 = C_2 = \frac{1}{4 \times \pi^2 \times 900^2\,\text{kHz} \times 200\,\mu\text{H}}$$

$$= 156.3\,\text{pF}$$

(b)   The required coefficient of coupling k is determined by f_0 and BW. Recall that BW $= kf_0$ for a double-tuned transformer. Solving for k gives us:

$$k = \frac{\text{BW}}{f_0}$$

$$= \frac{15\,\text{kHz}}{900\,\text{kHz}}$$

$$= 0.0167$$

(c)   In order to determine whether the circuit is undercoupled, overcoupled, or critically coupled, we must compare the value of k in Part (b) to the value of k_c for this circuit.

$$k_c = \frac{1}{\sqrt{Q_P Q_S}}$$

$$= \frac{1}{\sqrt{60 \times 75}}$$

$$= 0.0149$$

Since the value of k_c is slightly less than the value of k in Part (b), this circuit is said to be overcoupled.

Supplementary Problems

1.21   Calculate f_0 for a series LC circuit with $L = 125\,\mu\text{H}$ and $C = 10\,\text{pF}$.   *Ans.*   4.5 MHz

1.22   Calculate f_0 for a series LC circuit with $L = 250\,\mu\text{H}$ and $C = 20\,\text{pF}$.   *Ans.*   2.25 MHz

1.23   Calculate f_0 for a parallel LC circuit with $L = 400\,\mu\text{H}$ and $C = 200\,\text{pF}$.   *Ans.*   562.7 kHz

1.24   Calculate f_0 for a parallel LC circuit with $L = 4\,\mu\text{H}$ and $C = 28\,\text{pF}$.   *Ans.*   15 MHz

1.25   What value of inductance must be connected in series with a 120 pF capacitor to obtain an f_0 of 500 kHz?
Ans.   844.3 μH

1.26   What value of inductance must be connected in series with a 25 pF capacitor to obtain an f_0 of 5 MHz?
Ans.   40.5 μH

1.27   What value of capacitance must be connected in series with a 12 μH inductor to obtain an f_0 of 2.5 MHz?
Ans.   337.7 pF

1.28   What value of capacitance must be connected in parallel with a 2.5 μH inductor to obtain an f_0 of 10 MHz?
Ans.   101.3 pF

1.29   Refer to Fig. 1-21. With C set to 200 pF, calculate: (a) f_0, (b) X_L and X_C at f_0, (c) Z_T at f_0, (d) I at f_0, (e) Θ_Z at f_0,
(f) Q, (g) V_L and V_C at f_0, (h) BW, (i) f_1 and f_2 respectively.
Ans.   (a) 5 MHz, (b) 159 Ω, (c) 6.36 Ω, (d) 18.87 mA, (e) 0°, (f) 25, (g) 3 V, (h) 200 kHz, (i) 4.9 MHz, 5.1 MHz

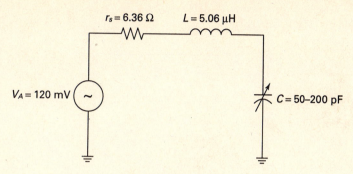

Fig. 1-21

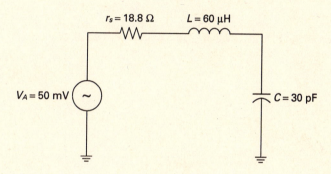

Fig. 1-22

1.30 Repeat Problem 1.29 with C set to 50 pF.
Ans. (a) 10 MHz, (b) 318 Ω ($X_L = X_C$), (c) 6.36 Ω, (d) 18.87 mA, (e) 0°, (f) 50, (g) 6 V, (h) 200 kHz, (i) 9.9 MHz, 10.1 MHz

1.31 In Fig. 1-21, assume that C is set to 50 pF. Calculate: (a) Z_T at f_1, (b) I at f_1, (c) Θ_Z at f_1, (d) Z_T at f_2, (e) I at f_2, (f) Θ_Z at f_2.
Ans. (a) 9 Ω, (b) 13.33 mA, (c) −45°, (d) 9 Ω, (e) 13.33 mA, (f) +45°

1.32 Refer to Fig. 1-22. Calculate the following: (a) f_0, (b) X_L and X_C at f_0, (c) Z_T at f_0, (d) I at f_0, (e) Q, (f) V_L and V_C at f_0, (g) BW, (h) f_1 and f_2 respectively.
Ans. (a) 3.75 MHz, (b) 1.41 kΩ, (c) 18.8 Ω, (d) 2.66 mA, (e) 75, (f) 3.75 V, (g) 50 kHz, (h) 3.725 MHz, 3.775 MHz

1.33 In Fig. 1-22, how much additional series resistance must be added to increase the bandwidth to 75 kHz?
Ans. 9.4 Ω

1.34 In Fig. 1-22, assume that the value of L is doubled and the value of C is halved. Calculate: (a) f_0, (b) X_L and X_C at f_0, (c) Q, (d) V_L and V_C at f_0, (e) BW.
Ans. (a) 3.75 MHz, (b) 2.83 kΩ, (c) 150.5, (d) 7.5 V, (e) 25 kHz

1.35 Refer to Fig. 1-23. Calculate the following: (a) f_0, (b) X_L and X_C at f_0, (c) Z_T at f_0, (d) I at f_0, (e) Q, (f) V_L and V_C at f_0, (g) BW, (h) f_1 and f_2 respectively.
Ans. (a) 830 kHz, (b) 234.68 Ω, (c) 2.83 Ω, (d) 3.53 mA, (e) 82.9 (f) 829 mV, (g) 10 kHz, (h) 825 kHz, 835 kHz

1.36 Refer to Fig. 1-23. Calculate the following: (a) Z_T at f_1 and f_2, (b) I at f_1 and f_2, (c) Θ_Z at f_1 and f_2 respectively, (d) BW if L is halved and C is doubled.
Ans. (a) 4 Ω, (b) 2.5 mA, (c) −45°, +45°, (d) 20 kHz

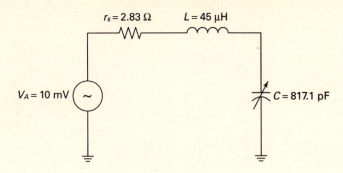

Fig. 1-23

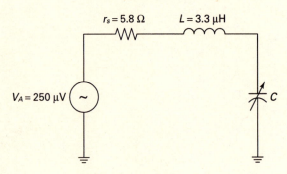

Fig. 1-24

1.37 Suppose that the series RLC circuit in Fig. 1-24 is to be designed so that it can tune over a frequency range of 14.0 MHz to 14.35 MHz. Calculate the following: (a) the required tuning range TR, (b) C_{min}, (c) C_{max}.
Ans. (a) TR = 1.05/1, (b) 37.28 pF, (c) 39.16 pF

1.38 Refer to Fig. 1-24. If C is adjusted to 38.07 pF, calculate: (a) f_0, (b) Z_T at f_0, (c) I at f_0, (d) Q, (e) V_L and V_C at f_0, (f) BW.
Ans. (a) 14.2 MHz, (b) 5.8 Ω, (c) 43.1 μA, (d) 50.76, (e) 12.69 mV, (f) 279.7 kHz

1.39 When either L or C is decreased in a series or parallel circuit LC circuit, what happens to the resonant frequency f_0? *Ans.* f_0 increases

1.40 Determine the amount of resistance contained within a series RLC circuit if $f_0 = 20$ MHz, $Q = 30$ and $L = 200\ \mu$H. *Ans.* 837.76 Ω

1.41 A 3 mH inductance in series with a capacitance is resonant at 1 MHz. If the coil has a Q of 40 at f_0, calculate: (a) the required value of C, (b) the coil resistance r_s, (c) BW.
Ans. (a) 8.44 pF, (b) 471.2 Ω, (c) 25 kHz

1.42 In Fig. 1-25, calculate the following: (a) f_0, (b) X_L and X_C at f_0, (c) I_L and I_c at f_0, (d) Q, (e) Z_{tank}, (f) I_T, (g) BW, (h) f_1 and f_2 respectively.
Ans. (a) 4.91 MHz, (b) 108 Ω, (c) 4.63 mA, (d) 108, (e) 11.66 kΩ, (f) 42.88 μA, (g) 45.46 kHz, (h) 4.89 MHz and 4.93 MHz

1.43 In Fig. 1-25, assume that a 5 kΩ load has been placed in parallel with the tank. Calculate the following: (a) Q_{ckt}, (b) BW, (c) total line current I_T.
Ans. (a) 32.4, (b) 151.55 kHz, (c) 142.86 μA

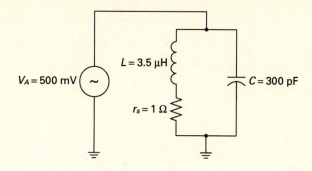

Fig. 1-25

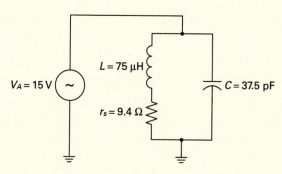

Fig. 1-26

1.44   Refer to Fig. 1-26. Calculate the following: (a) f_0, (b) X_L and X_C at f_0, (c) I_L and I_C at f_0, (d) Q, (e) Z_{tank}, (f) I_T, (g) BW, (h) f_1 and f_2 respectively.
    Ans.   (a) 3 MHz, (b) 1.41 kΩ, (c) 10.64 mA, (d) 150, (e) 211.5 kΩ, (f) 70.92 μA, (g) 20 kHz, (h) 2.99 MHz, 3.01 MHz

1.45   In Fig. 1-26, suppose that C is increased to 75 pF. Calculate: (a) f_0, (b) X_L and X_C at f_0, (c) I_L and I_C at f_0, (d) Q, (e) Z_{tank}, (f) I_T, (g) BW.
    Ans.   (a) 2.12 MHz, (b) 999 Ω, (c) 15 mA, (d) 106.3, (e) 106.3 kΩ, (f) 141.1 μA, (g) 19.94 kHz

1.46   Refer to Fig. 1-26. Assume that a 50 kΩ load is placed in parallel with the tank. Calculate the following ($C = 37.5$ pF): (a) Q_{ckt}, (b) BW, (c) I_T.
    Ans.   (a) 28.7, (b) 104.5 kHz, (c) 371 μA

1.47   Refer to Fig. 1-27. Calculate the following (S_1 is open): (a) f_0, (b) X_L and X_C at f_0, (c) I_L and I_C at f_0, (d) Q, (e) Z_{tank} at f_0, (f) I_T at f_0.
    Ans.   (a) 1 MHz, (b) 1.5 kΩ, (c) 10 mA, (d) 150, (e) 225 kΩ, (f) 66.67 μA

1.48   In Fig. 1-27, calculate the following: (a) BW with S_1 open, (b) BW with S_1 closed.
    Ans.   (a) 6.67 kHz, (b) 100 kHz

1.49   Refer to Fig. 1-27. Determine whether the tank appears inductive, capacitive or resistive at each of the following frequencies: (a) f_1, (b) f_2, (c) f_0.
    Ans.   (a) inductive, (b) capacitive, (c) resistive

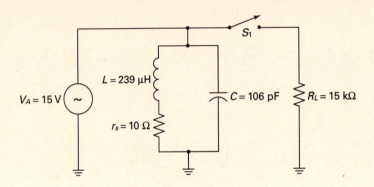

Fig. 1-27

1.50 Refer to Fig. 1-27. Determine the phase relationship between I_L and I_C at each of the following frequencies (ignore r_s): (a) f_0, (b) f_1, (c) f_2.
Ans. (a) 180°, (b) 180°, (c) 180°

1.51 Determine the necessary coefficient of coupling in order to provide critical coupling for a double-tuned transformer in which $Q_P = 25$ and $Q_S = 40$.   *Ans.* 0.0316

1.52 The specifications for a double-tuned transformer to be used as a coupling network are such as to require a resonant frequency of 700 kHz and a bandwidth of 25 kHz. Both the primary and secondary of the transformer have an inductance of 75 μH. The primary circuit has a Q of 20 while the secondary has a Q of 15.

(a) Determine the capacitance required across each of the coils.

(b) What coefficient of coupling is required in order to meet these specifications?

(c) Is this circuit undercoupled, critically coupled, or overcoupled?

Ans. (a) 689.2 pF, (b) 0.0357, (c) undercoupled

Chapter 2

RF Oscillators, PLLs, and Frequency Synthesizers

INTRODUCTION

Radio-frequency (RF) oscillators play a very important role in every electronic communication system. By definition, an oscillator is a circuit capable of continuously generating a repetitive waveform of the desired frequency. Depending on the application, the repetitive waveform may be either sinusoidal or rectangular in nature. If an oscillator requires no external input signal or trigger to produce the desired output waveform, it is said to be self-sustaining or free-running. Those oscillators requiring an external input signal are called triggered or one-shot oscillators. In this chapter, we will deal only with self-sustaining or free-running oscillators. Also included in this chapter is the coverage of frequency synthesizers which are special frequency-generating circuits encountered in the most modern electronic communication systems. Like the oscillator, a frequency synthesizer produces a repetitive waveform of the desired frequency. Because a frequency synthesizer uses a phase-locked loop (PLL) as a basic building block, complete coverage of phase-locked loops is provided prior to the coverage of frequency synthesizers.

2.1 OSCILLATOR FUNDAMENTALS

Figure 2-1(*a*) shows the block diagram of an oscillator containing an amplifier and a feedback network. The output voltage V_{out} from the amplifier is calculated as:

$$V_{\text{out}} = V_{\text{in}} \times A_V \tag{2.1}$$

where V_{in} represents the total input voltage applied to the amplifier.

A portion of the output voltage is fed back to the input of the amplifier through the feedback network. The fractional part of the output voltage which is fed back to the input is called the feedback voltage and is designated V_{fb}. The ratio of feedback voltage V_{fb} to output voltage V_{out} is called the feedback fraction and is designated B. The feedback fraction can be expressed as:

$$B = \frac{V_{\text{fb}}}{V_{\text{out}}} \tag{2.2}$$

Rearranging Eq. (*2.2*) gives us:

$$V_{\text{fb}} = BV_{\text{out}} \tag{2.3}$$

Equation (*2.3*) clearly states that the feedback voltage is a fractional part of the output voltage. As you can see in Fig. 2-1(*a*), the feedback voltage V_{fb} is in series with the source voltage V_S. If the phase of the feedback voltage is negative, then a portion of V_S will be cancelled, in which case the output voltage will be reduced in amplitude. If the phase of the feedback voltage is positive, however, then the input voltage V_{in} is larger than with V_S alone. All oscillators must use positive feedback to sustain oscillations. Assuming the feedback voltage is positive in Fig. 2-1(*a*), the actual input voltage to the amplifier is:

$$V_{\text{in}} = V_S + V_{\text{fb}} \tag{2.4}$$

Substituting $V_S + V_{\text{fb}}$ for V_{in} in Eq. (*2.1*) gives us:

$$V_{\text{out}} = (V_S + V_{\text{fb}})A_V \tag{2.5}$$

Replacing V_{fb} with BV_{out} gives us:

$$V_{\text{out}} = (V_S + BV_{\text{out}})A_V \tag{2.6}$$

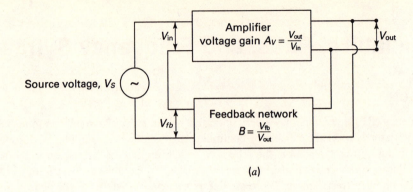

(a)

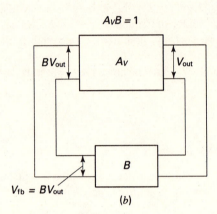

(b)

Fig. 2-1

or

$$V_{out} = A_V V_S + A_V B V_{out} \qquad (2.7)$$

Take a close look at Eqs. (2.6) and (2.7). They will be true only if two conditions are met. V_S must equal zero and the $A_V B$ product must equal 1. The requirement for the $A_V B$ product to equal 1 is called the *Barkhausen criterion*. For any practical oscillator, the $A_V B$ product may equal 1 or more. Essentially, the Barkhausen criterion states that the amplifier voltage gain must be at least large enough to overcome the losses in the feedback network. In equation form, the Barkhausen criterion is expressed as:

$$A_V B = 1 \qquad (2.8)$$

If $A_V B < 1$, the oscillator will not function.

Figure 2-1(*a*) does not illustrate a true oscillator, because of the presence of the external voltage source V_S. The only input to a real self-sustaining or free-running oscillator is the dc power supply from which power is supplied to the circuit. Figure 2-1(*b*) shows the block diagram of a real oscillator. As long as the Barkhausen criterion is met ($A_V B = 1$), the output voltage V_{out} is developed without an external voltage source V_S. The feedback voltage $B V_{out}$ serves as the only input to the amplifier.

If the amplifier portion of an oscillator provides a 180° phase shift between its input and output signals, then A_V is represented as a negative quantity. An example would be a common-emitter amplifier where $A_V = -V_{out}/V_{in}$. With A_V negative, the feedback fraction B must also be negative in order to satisfy the Barkhausen criterion. In other words, if the amplifier portion of an oscillator provides a 180° phase shift, the feedback network must also provide a 180° phase shift.

In summary, an oscillator will be self-sustaining if its $A_V B$ product equals 1. If $A_V B < 1$ or if the $A_V B$ product is negative, then the circuit will not oscillate.

The following is a list of the four requirements required for a self-sustaining oscillator to work.

1.   *A dc power source.* There must be a source of electrical power: either a dc power supply or a battery.

2.   *Frequency-determining components.* There must be a component or components which control the frequency of oscillation. In some cases, the oscillator may operate at only a single frequency, whereas in other cases it may need to be variable over a given range of frequencies.

3.   *Amplification.* The oscillator circuit must be capable of amplification. The gain of the amplifier must be large enough to overcome the losses in the feedback network. In other words, the $A_V B$ product must equal 1.

4.   *Positive feedback.* There must be a path for the output signal to be fed back to the input. The feedback signal must have the correct phase and amplitude to sustain oscillations. If the phase of the feedback voltage is incorrect or if the feedback voltage is too small, oscillations will cease. The term positive feedback refers to the fact that the signal fed back from output to input aids or reinforces the original input signal.

EXAMPLE 2.1.   In Fig. 2-1(*b*), assume that the feedback fraction B equals 0.01. Calculate the required value of A_V in order to sustain oscillations.

   Ans.   The Barkhausen criterion states that in order for an oscillator to be self-sustaining, the $A_V B$ product must equal 1. Therefore,

$$A_V B = 1$$

Solving for A_V gives us:

$$A_V = \frac{1}{B}$$

Inserting 0.01 for the value of B gives us:

$$A_V = \frac{1}{0.01}$$
$$= 100$$

This means that the voltage gain A_V must be at least 100 in order to sustain oscillations. If A_V should decrease below 100, then oscillations will cease.

EXAMPLE 2.2.   In Fig. 2-1(*b*), assume that $A_V = -100$. Solve for the value of B.

   Ans.   Again, recall that $A_V B$ must equal 1. Inserting the value of A_V and solving for B gives us:

$$B = \frac{1}{A_V}$$
$$= \frac{1}{-100}$$
$$= -0.01$$

The negative quantity of -0.01 for B indicates that a 180° phase shift must be provided by the feedback network.

2.2   RF OSCILLATOR CIRCUITS

Let us apply our knowledge of oscillator fundamentals to some commonly used oscillator circuits. Figure 2-2(*a*) shows a popular circuit known as a *Hartley oscillator*. Its identifiable feature is the tapped coil in the LC tank circuit. Notice that the coil's tap point is grounded.

The amplifier portion of the oscillator is a common-emitter amplifier. Therefore, the ac base voltage and ac collector voltage are 180° out of phase. This means that the feedback network must provide an additional 180° phase shift so that the feedback voltage is of the correct phase to sustain oscillations. To help you understand the circuit's operation, it has been redrawn in Fig. 2-2(*b*). Notice that L_1 is in parallel with L_2 and C_3 in series. At the resonant frequency of the tank circuit, the net reactance of the $L_2 C_3$ branch

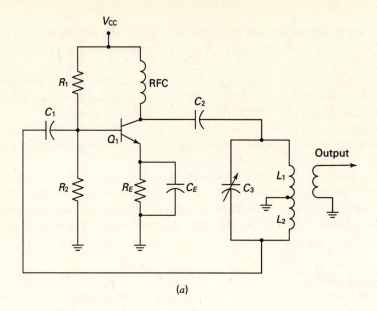

(a)

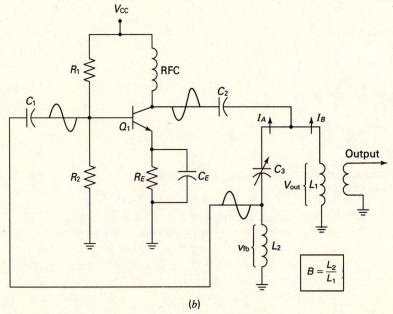

(b)

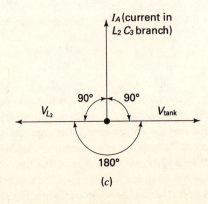

(c)

Fig. 2-2

appears capacitive. Also, the net capacitive reactance of the $L_2 C_3$ branch equals the inductive reactance of the L_1 branch. This gives us the following mathematical relationship.

$$X_{L_1} = X_{C_3} - X_{L_2}$$

Rearranging terms we have:

$$X_{L_1} + X_{L_2} = X_{C_3}$$

Therefore, the frequency of oscillation, labeled f_{osc}, equals the frequency at which $X_{L_1} + X_{L_2} = X_{C_3}$.

$$f_{osc} = \frac{1}{2\pi\sqrt{(L_1 + L_2)C_3}} \tag{2.9}$$

As you can see in Fig. 2-2(b), the feedback voltage v_{fb} driving the base of the transistor is developed across the inductance L_2. Similarly, the output voltage is developed across the inductance L_1. As you recall from Section 2-1, the ratio of feedback voltage to output voltage is the feedback fraction B. In Fig. 2-2(b), B equals:

$$B = \frac{v_{fb}}{v_{out}}$$

$$= \frac{I \times X_{L_2}}{I \times X_{L_1}} = \frac{2\pi f L_2}{2\pi f L_1}$$

or

$$B = \frac{L_2}{L_1} \tag{2.10}$$

To satisfy the Barkhausen criterion for oscillation, the $A_V B$ product must equal 1. Solving for A_V gives us:

$$A_V = \frac{1}{B}$$

or

$$A_V = \frac{L_1}{L_2} \tag{2.11}$$

This says that the voltage gain of the common-emitter amplifier must equal at least the ratio L_1/L_2 in order to overcome the losses in the feedback network.

Next, let us analyze how the feedback voltage is shifted 180° in phase by the feedback network. See the phasor diagram in Fig. 2-2(c). The ac voltage across the tank is in phase with the net line current I_T of the tank, since the tank appears as a very large resistance at the resonant frequency. Therefore, V_{tank} is used as the reference phasor in Fig. 2-2(c). Since the net reactance of the $L_2 C_3$ branch is capacitive, the current I_A in this branch leads V_{tank} by 90° as shown. Since the voltage across an inductance leads its current by 90°, and since I_A also flows through L_2, V_{L_2} must lead I_A by 90°. Therefore V_{L_2} is shown to lead I_A by 90° in Fig. 2-2(c). It is clear therefore that the ac voltage developed across L_2 is 180° out of phase with the ac voltage developed across the tank. This is how simple it is. Starting at the base and moving to the collector, there is a 180° phase shift. Through the feedback network there is another 180° phase shift. Therefore, moving from base to collector and back to the base, there is a total phase shift of 360°.

To vary the frequency of oscillation in a Hartley oscillator, the capacitance of the tank circuit is varied, rather than the inductances L_1 and L_2. The reason is that it is very difficult to vary L_1 and L_2 while still maintaining the proper feedback fraction B. In Fig. 2-2(a), the tank capacitor C_3 is shown to be a variable capacitance. The output from a Hartley oscillator is usually coupled to the next circuit by a transformer. In some cases however, the output may be capacitively coupled.

In Fig. 2-2(a), the capacitors C_1 and C_2 act as coupling capacitors which block dc. If the feedback voltage were directly coupled to the base of Q_1, the dc voltage at the base would be shorted out by the low dc resistance of L_2.

Similarly, if the ac collector voltage were directly coupled to the tank, L_1 would short the dc collector voltage. The radio-frequency choke (RFC) in series with the collector and V_{CC} presents a very high impedance to ac signals. This effectively isolates the transistor collector from ac ground. The dc resistance of the RFC is negligible, however. The resistors R_1, R_2, and R_E provide the circuit with the proper dc bias.

EXAMPLE 2.3. Refer to Fig. 2-2(a). Assume that it is desired to design the oscillator so that its oscillating frequency is 4 MHz. If C_3 is adjusted 300 pF and the desired feedback fraction is $B = \frac{1}{20}$, calculate: (a) the required value of total tank inductance L_T ($L_T = L_1 + L_2$), (b) the required values for L_1 and L_2, (c) the minimum voltage gain A_V required of the common emitter amplifier to sustain oscillations.

  Ans.

(a) Recall that f_{osc} is calculated using the following equation:

$$f_{osc} = \frac{1}{2\pi\sqrt{(L_1 + L_2)C_3}}$$

Solving for $L_1 + L_2$ gives us:

$$L_1 + L_2 = \frac{1}{4\pi^2 f_{osc}^2 C_3}$$

$$= \frac{1}{4 \times \pi^2 \times 4^2\,\text{MHz} \times 300\,\text{pF}}$$

$$= 5.28\,\mu\text{H}$$

(The sum $L_1 + L_2$ can be represented as L_T for convenience.)

(b) In Part (a), $L_T = 5.28\,\mu\text{H}$. Recall that the feedback fraction B in Fig. 2-2(a) is equal to:

$$B = \frac{L_2}{L_1}$$

Since it is desired to have $B = \frac{1}{20}$, the following relationship is true.

$$\frac{L_2}{L_1} = \frac{1}{20}$$

Stating L_1 in terms of L_2 gives us:

$$L_1 = 20L_2$$

This means that the total inductance L_T can be expressed as:

$$L_T = L_1 + L_2$$
$$= 20L_2 + L_2$$
$$L_T = 21L_2$$

Solving for L_2 gives us:

$$L_2 = \frac{L_T}{21}$$

$$= \frac{5.28\,\mu\text{H}}{21}$$

$$= 0.251\,\mu\text{H}$$

To find L_1, begin with $L_T = L_1 + L_2$. Next, solve for L_1:

$$L_1 = L_T - L_2$$
$$= 5.28\,\mu\text{H} - 0.251\,\mu\text{H}$$
$$= 5.029\,\mu\text{H}$$

(c) To meet the Barkhausen criterion, recall that $A_V B$ must equal 1. Therefore,

$$A_V = \frac{1}{B}$$

or

$$A_V = \frac{L_1}{L_2}$$

$$= \frac{20}{1}$$

$$A_V = 20$$

This means that the common-emitter amplifier must have a voltage gain of at least 20 to sustain oscillations.

Colpitts Oscillator

Another common RF oscillator is the *Colpitts oscillator* shown in Fig. 2-3(a). Its identifiable feature is the tapped capacitors in the LC tank circuit. Like the Hartley oscillator, the tap point is grounded. The amplifier portion of the Colpitts oscillator in Fig. 2-3(a) is a common-emitter amplifier. Since the common-emitter amplifier provides a 180° phase shift between the base and collector, the feedback network must provide an additional 180° phase shift. To help understand circuit operation, refer to the equivalent circuit in Fig. 2-3(b). Notice that C_1 is in parallel with L_1 and C_2 in series. At the resonant frequency of the tank circuit, the net reactance of the $L_1 C_2$ branch appears inductive. Also, the net inductive reactance of the $L_1 C_2$ branch equals the capacitive reactance of the C_1 branch. This gives us the following mathematical relationship:

$$X_{C_1} = X_{L_1} - X_{C_2}$$

Rearranging terms, we have:

$$X_{L_1} = X_{C_1} + X_{C_2}$$

Therefore, the frequency of oscillation equals the frequency at which $X_{L_1} = X_{C_1} + X_{C_2}$:

$$f_{osc} = \frac{1}{2\pi\sqrt{L_1 C_{eq}}} \tag{2.12}$$

where

$$C_{eq} = \frac{C_1 C_2}{C_1 + C_2}$$

In Fig. 2-3(b), the feedback voltage is developed across C_2 in the tank circuit. Similarly, the output voltage is developed across C_1. This means the feedback fraction B equals:

$$B = \frac{v_{fb}}{v_{out}}$$

$$= \frac{IX_{C_2}}{IX_{C_1}} = \frac{X_{C_2}}{X_{C_1}} = \frac{\dfrac{1}{2\pi f C_2}}{\dfrac{1}{2\pi f C_1}}$$

or

$$B = \frac{C_1}{C_2} \tag{2.13}$$

To satisfy the Barkhausen criterion for oscillations, $A_V B = 1$. Solving for A_V gives us:

$$A_V = \frac{1}{B}$$

or

$$A_V = \frac{C_2}{C_1} \tag{2.14}$$

Therefore, in order to sustain oscillations in a Colpitts oscillator, the common-emitter amplifier must have a voltage gain equal to or greater than the ratio C_2/C_1.

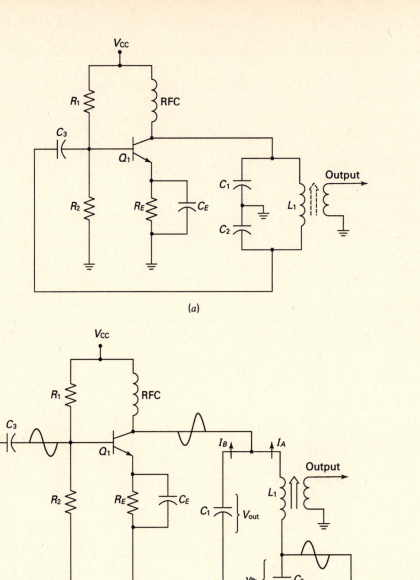

(a)

(b)

(c)

Fig. 2-3

Next, let us analyze how the feedback voltage is shifted 180° in phase by the feedback network. See the phasor diagram in Fig. 2-3(c). The ac voltage across the tank is in phase with the net line current I_T of the tank, since the tank appears as a very large resistance at the resonant frequency. Therefore, V_{tank} is used as the reference phasor in Fig. 2-3(c). Since the net reactance of the L_1C_2 branch is inductive at resonance, the branch current I_A lags V_{tank} by 90° as shown. Also, since the voltage across a capacitance lags the current flowing to and from its plates by 90° and since I_A also flows through C_2, V_{C_2} is shown lagging I_A by 90° in Fig. 2-3(c). Therefore, it can be seen that the ac voltage developed across C_2 lags V_{tank} by 180°. Moving from base to collector, through the feedback network and back to the base, there is a total phase shift of 360°.

To vary the frequency of oscillation in a Colpitts oscillator, the inductance L_1 is varied rather than C_1 or C_2. The reason is that varying L_1 will not vary the feedback fraction B. In Fig. 2-3(a), L_1 is usually adjusted by moving a ferrite slug in and out of its core.

In Fig. 2-3(a), R_1, R_2, R_E, and RFC serve the same purpose as in the Hartley oscillator of Fig. 2-2(a). Note, however, that the coupling capacitor between the collector and tank circuit is no longer necessary. This is because C_1 appears as an open to dc and C_3 isolates the dc collector and base voltages.

EXAMPLE 2.4.   Refer to Fig. 2-3(a). Assume that it is desired to design the oscillator so that its oscillating frequency is 4 MHz. If $L_1 = 50\ \mu H$ and the feedback fraction $B = \frac{1}{20}$, calculate: (a) the required value of equivalent tank capacitance ($1/C_{eq} = 1/C_1 + 1/C_2$), (b) the required values of C_1 and C_2, (c) the minimum voltage gain A_V required of the common-emitter amplifier to sustain oscillations.

   Ans.

(a)   Recall that f_{osc} is calculated using the following equation.

$$f_{osc} = \frac{1}{2\pi\sqrt{LC_{eq}}}$$

Solving for C_{eq} gives us:

$$C_{eq} = \frac{1}{4\pi^2 f_{osc}^2 L_1}$$

$$= \frac{1}{4 \times \pi^2 \times 4^2\,\text{MHz} \times 50\,\mu H}$$

$$= 31.66\,\text{pF}$$

(b)   Recall in Fig. 2-3(a) that $B = C_1/C_2$. Therefore the following is true.

$$\frac{1}{20} = \frac{C_1}{C_2}$$

Solving for C_2 gives us:

$$C_2 = 20C_1$$

This allows us to state C_{eq} as:

$$\frac{1}{C_{eq}} = \frac{1}{C_1} + \frac{1}{20C_1}$$

which simplifies to:

$$\frac{1}{C_{eq}} = \frac{21}{20C_1}$$

Solving for C_1 gives us:

$$C_1 = C_{eq} \times \frac{21}{20}$$

$$= 31.66\,\text{pF} \times \frac{21}{20}$$

$$= 33.24\,\text{pF}$$

Solving for C_2 gives us:

$$\frac{1}{C_{eq}} = \frac{1}{C_1} + \frac{1}{C_2}$$

$$C_2 = \frac{1}{\dfrac{1}{C_{eq}} - \dfrac{1}{C_1}}$$

$$= \frac{1}{\dfrac{1}{31.66\,\text{pF}} - \dfrac{1}{33.24\,\text{pF}}}$$

$$= 666.06\,\text{pF}$$

(c)   To meet the Barkhausen criterion, recall that $A_V B = 1$. Therefore,

$$A_V = \frac{1}{B}$$

or

$$A_V = \frac{C_2}{C_1}$$

$$= \frac{20}{1}$$

So

$$A_V = 20$$

Oscillator Start-up

You may be wondering how oscillators such as those shown in Figs. 2-2 and 2-3 get started, since there is no external signal initially applied to get the oscillations going. Where does the initial input signal come from? When power is first applied, the sudden burst of dc current contains frequencies from 0 Hz to well over 1 GHz. These frequency components (small ac voltages) are amplified by the transistor and appear at the collector. Also, a certain portion of the output voltage is fed back to the input via the feedback network. The common-emitter amplifier provides a very large voltage gain only at the tank circuit's resonant frequency. Therefore only the ac noise voltage which has a frequency equal to the resonant frequency of the tank circuit, is amplified by a considerable factor. This signal voltage is then fed back a little stronger at the base than it initially appeared at the time the dc power was first applied. The favored resonant frequency continues to build up until it reaches its maximum possible value. This brings us to a very important point. In order for any self-sustaining oscillator to start up, the $A_V B$ product, also called the circuit's *loop gain*, must initially be greater than 1. When oscillations are first building up, the amplifier portion of the oscillator operates in the small-signal mode with maximum voltage gain. As the output signal amplitude builds up, because of positive feedback, the amplifier voltage gain decreases. The reason is that the amplifier reaches its limits in terms of its maximum output voltage. Eventually the loop gain $A_V B$ decreases to unity (1) as oscillations stabilize.

It should be noted that the output waveform of most RF oscillators is not a perfect sine wave. However, the LC tank circuit helps eliminate most of the frequency components that distort the fundamental-frequency sine wave.

Tuned-input Tuned-output Oscillator

Figure 2-4(a) shows another type of oscillator, known as a tuned-input tuned-output oscillator. L_1 and C_1 make up the tuned circuit on the base or input side whereas L_2 and C_2 make up the tuned circuit on the output or collector side. The component that makes the circuit oscillate, however, is not what you might expect: it is the collector-base capacitance C_{cb}, which is shown with the dashed lines in Fig. 2-4(a). This

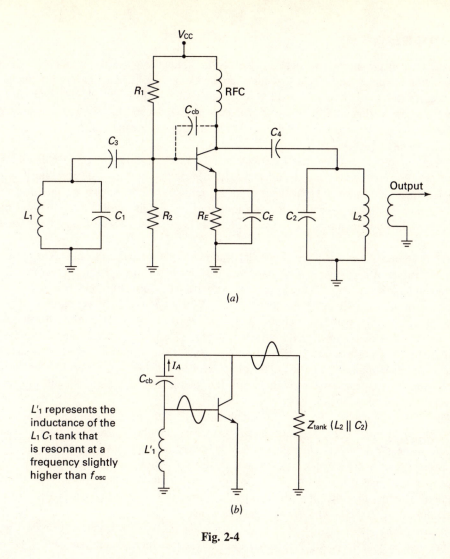

Fig. 2-4

is the interelectrode capacitance within the transistor. Usually, C_{cb} is just a few picofarads or so, with 1 to 2 pF being very typical. The capacitance C_{cb} is the component which provides the positive feedback from the output back to the input. Because the ac signal at the base is considered to be the input signal, and because the output signal is at the collector, the amplifier portion of the oscillator is a common-emitter amplifier. Because of this, the feedback network must provide an additional 180° phase shift from collector to base. To see how the 180° phase shift is accomplished, refer to the equivalent circuit in Fig. 2-4(b).

The input tuned circuit, consisting of L_1 and C_1, is tuned to a resonant frequency slightly higher than the oscillating frequency f_{osc}, which is determined by L_2 and C_2. As a result, the $L_1 C_1$ tank appears as an inductance L_1' since it is operating below its natural resonant frequency. Next, consider that C_{cb} has an X_C that is very large due to its very small capacitance value. Therefore $X_{cb} \gg X_{L_1'}$, and the series combination of C_{cb} and L_1' has a net reactance that is capacitive. Therefore, the current I_A in this branch leads the voltage at the collector by 90°. Since the voltage across L_1' must lead its current by 90°, $V_{L_1'}$ must lead the ac collector voltage by 180°. Therefore, the circuit has a feedback voltage of the correct phase to sustain oscillations.

In some cases, it may be necessary to add a very small capacitance across the base and collector of the transistor. This is done to provide sufficient feedback to start and sustain oscillations.

Crystal Oscillators

In today's electronic communication systems, the frequency stability of an oscillator is of primary concern. For example, the Federal Communications Commission (FCC) has very stringent regulations regarding the stability of both AM and FM broadcast stations. FM broadcast stations must maintain their carrier frequency within ± 2 kHz of the assigned carrier frequency, which is approximately a tolerance of $\pm 0.002\%$. In the AM broadcast band, the maximum allowable carrier drift is only ± 20 Hz from the assigned carrier frequency.

There are several different factors affecting the frequency stability of an oscillator, with the most obvious being those which directly affect the values of the frequency-determining components. For example, if the inductance or capacitance of an LC circuit varies with changes in either temperature or humidity, then it will cause a shift in the operating frequency of the oscillator. To improve the stability of an oscillator, it is common practice to use a crystal as the frequency-determining component rather than an LC circuit.

The operation of a crystal is based on what is known as the piezoelectric effect. The piezoelectric effect is the generation of a voltage in a crystal when it is subjected to mechanical pressure. The three most prominent piezoelectric crystals used today are rochelle salt, tourmaline, and quartz. Of the three crystal types mentioned, quartz is by far the most popular. When a crystal such as quartz is subjected to an ac voltage, it alternately expands and contracts, which means that it vibrates. It has a natural frequency of vibration which is determined by its physical dimensions and its operating temperature. A raw quartz crystal is cut into thin slices by a high-precision saw. Such a slice is then mounted between two conducting plates with connecting leads attached to the plates. If the temperature of the crystal slice is held constant, then its vibrating frequency will be highly stable. For this reason, crystals find application in transmitters, receivers, wristwatches, etc.

The schematic symbol of a crystal is shown in Fig. 2-5(a). Its electrical equivalent circuit is shown in Fig. 2-5(b). In Fig. 2-5(b), the inductance L represents the vibrating mass, C the mechanical stiffness, and R the mechanical friction. The values of L and C determine the series resonant frequency of the crystal. The crystal's socket has a capacitance which exists in parallel with the crystal itself. This mounting capacitance, designated C_M, forms a parallel resonant circuit with the crystal. The Q of a crystal is considerably higher than that of a conventional LC circuit. In fact, the Q of a typical crystal may range from about 10,000 to 500,000 in some special designs.

Figure 2-5(c) shows a graph of reactance versus frequency for a typical quartz crystal. The series resonant frequency of the crystal is represented as f_s. The parallel resonant frequency of the crystal and its holder is designated f_p. The frequencies f_s and f_p are separated by only a very slight amount. In Fig. 2-5(c), note that below f_s the crystal exhibits a capacitive reactance X_C, and above f_s it exhibits an inductive reactance X_L. At f_s, the impedance of the crystal is extremely low. Above the series resonant frequency f_s, the crystal acts like an inductance and therefore forms a parallel resonant circuit with the socket or mounting capacitance C_M. At f_p, the crystal exhibits nearly infinite impedance.

Now let us take a look at some typical crystal-oscillator circuits. Figure 2-6 shows several different variations. Figure 2-6(a) shows a crystal-controlled Hartley oscillator, whereas Fig. 2-6(b) shows a crystal-controlled Colpitts oscillator. For these two circuits, the oscillating frequency is the same as the series resonant frequency f_s of the crystal. The reason is that, at f_s, the crystal is effectively a short circuit, thereby allowing the feedback voltage to be fed back to the base of the transistor. Slightly above and below f_s, the impedance of the crystal increases sharply, and the amplitude of the feedback voltage is insufficient to sustain oscillations. Also, above and below f_s, the reactance of the crystal causes the phase angle of the feedback voltage to be either greater or less than the 180° required for positive feedback. Since the crystal's Q is extremely high, the oscillating frequency of the circuits shown in Figs. 2-6(a) and 2-6(b) is extremely close to the series resonant frequency f_s of the crystal.

Figure 2-6(c) shows another variation of a Colpitts oscillator. In this case, the crystal Y_1 replaces the inductor in the tank circuit. To make the circuit oscillate however, the crystal must operate on the inductive side (high side) of series resonance (f_s). With Y_1 acting as an inductance, the feedback voltage across C_2 is 180° out of phase with the ac voltage present across the tank.

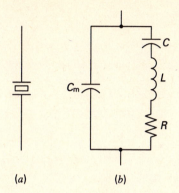

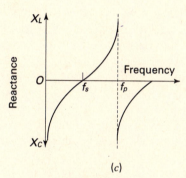

Fig. 2-5

The oscillator shown in Fig. 2-6(*d*) is known as a *Pierce oscillator*. For this oscillator, the crystal Y_1 operates at its parallel resonant frequency f_p. The voltage across C_2 serves as the feedback voltage which drives the emitter of the transistor. The bypass capacitor C_4 places the base at ground for ac signals. Since the output is taken from the collector and the input is applied to the emitter, the amplifier portion of the oscillator is a common-base amplifier. (As you recall, the input and output signals of a common-base amplifier are in phase.) When the ac collector voltage goes positive, the feedback voltage driving the emitter also goes positive. The positive-going signal at the emitter forces the collector voltage to go increasingly more positive. This continues until the collector voltage reaches its maximum value. When the ac collector voltage goes negative, the emitter voltage is also driven negative, causing the collector to become increasingly more negative. As you can see, the feedback voltage driving the emitter reinforces the original input signal, thus making the feedback positive.

EXAMPLE 2.5.   In Fig. 2-6(*d*), the feedback fraction B equals $\frac{1}{25}$. If C_1 and C_2 have an equivalent capacitance C_{eq} of 250 pF, calculate the values for C_1 and C_2.

Ans.   In Fig. 2-6(*d*) the feedback fraction B equals the ratio C_{eq}/C_2. This is verified in the following mathematical proof.

The ac current through C_1 and C_2 is identified as I. Since the output voltage is developed across the series combination of C_1 and C_2, the voltage v_{out} can be specified as:

$$v_{\text{out}} = IX_{C_{eq}}$$

The feedback voltage across C_2 equals:

$$v_{\text{fb}} = IX_{C_2}$$

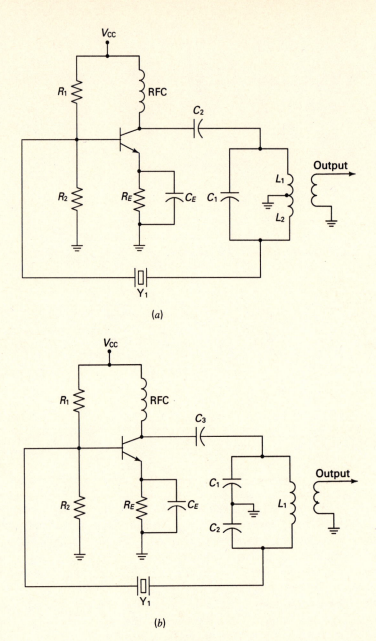

Fig. 2-6

Since $B = v_{fb}/v_{out}$, we have:

$$B = \frac{v_{fb}}{v_{out}}$$

$$= \frac{IX_{C_2}}{IX_{C_{eq}}}$$

or

$$B = \frac{X_{C_2}}{X_{C_{eq}}}$$

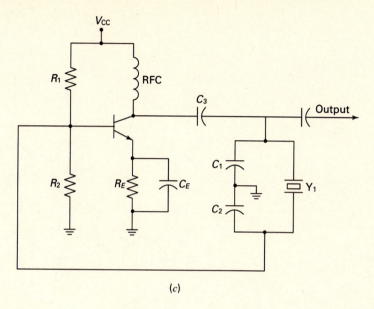

(c)

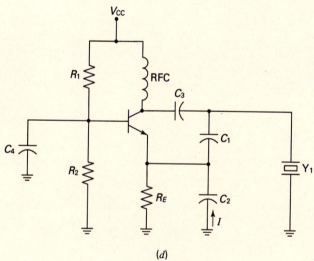

(d)

Fig. 2-6 (continued)

Expanding X_{C_2} to $1/2\pi f C_2$ and $X_{C_{eq}}$ to $1/2\pi f C_{eq}$ gives us

$$B = \frac{\dfrac{1}{2\pi f C_2}}{\dfrac{1}{2\pi f C_{eq}}}$$

or

$$B = \frac{C_{eq}}{C_2}$$

Therefore, in Fig. 2-6(d) we have:

$$\frac{1}{25} = \frac{C_{eq}}{C_2}$$

Rearranging to solve for C_2 gives:

$$C_2 = 25C_{eq}$$

Since $C_{eq} = 250$ pF, C_2 equals:

$$C_2 = 25 \times 250\,\text{pF}$$
$$= 6.25\,\text{nF}$$

Solving for C_1 gives:

$$C_1 = \cfrac{1}{\cfrac{1}{C_{eq}} - \cfrac{1}{C_2}}$$
$$= \cfrac{1}{\cfrac{1}{250\,\text{pF}} - \cfrac{1}{6.25\,\text{nF}}}$$
$$= 260.41\,\text{pF}$$

2.3 PHASE-LOCKED LOOP

A phase-locked loop (PLL) is basically an electronic feedback loop which consists of a phase detector (also called a phase comparator), a low-pass filter, a dc amplifier, and a voltage-controlled oscillator (VCO). The basic block diagram of a PLL is shown in Fig. 2-7. The functional blocks between the input and output are considered to be in the forward path of the loop, whereas the single connection between the VCO and phase detector is the feedback path. Before examining the overall operation of a PLL, let us examine each block separately.

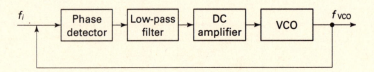

Fig. 2-7

VCO

A voltage-controlled oscillator is a free-running oscillator whose frequency of operation is controlled by an external dc bias voltage. In essence therefore, the input to a VCO is a dc voltage and the output is a frequency. Figure 2-8 shows the transfer curve for a typical VCO, which is a graph of the VCO's output frequency versus its dc input voltage. The output frequency with a dc input of 0 V is called the VCO's *natural frequency* and is designated f_N. Any change in the output frequency caused by a change in the dc input voltage is called *frequency deviation* and is designated Δf. The transfer function or conversion gain K of the VCO can be expressed as the amount of frequency deviation Δf per unit change Δv in dc input voltage.

Expressed mathematically,

$$K = \frac{\Delta f}{\Delta V} \tag{2.15}$$

where K is the input/output transfer function specified in Hz/V, $\Delta V =$ change in input voltage, and $\Delta f =$ frequency deviation.

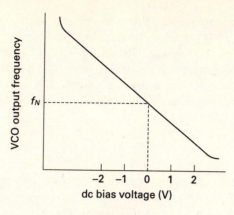

Fig. 2-8

In Fig. 2-8, it can be seen that the output frequency of the VCO decreases below the frequency f_N when the dc input voltage becomes increasingly positive. Conversely, the output frequency increases above f_N when the dc input voltage becomes increasingly negative. Depending on the exact amount of dc input voltage, the VCO's frequency of operation can be stated as $f_{VCO} = f_N \pm \Delta f$. When the frequency deviation Δf is symmetrical, the VCO's natural frequency f_N is centered within the linear portion of the transfer curve.

EXAMPLE 2.6.   The output frequency of a VCO changes from 100 kHz to 150 kHz when a change of 0.5 V occurs at the VCO input. Calculate the conversion gain K.

   Ans.

$$K = \frac{\Delta f}{\Delta V}$$

$$= \frac{150\,\text{kHz} - 100\,\text{kHz}}{0.5\,\text{V}}$$

$$= 100\,\text{kHz/V}$$

Phase Detector

A phase detector is a nonlinear circuit with two input signals. Referring to Fig. 2-7, we see that one of the input signals to the phase detector is an external signal f_i, and the other is the VCO output signal f_{VCO}. The output from the phase detector is a dc voltage whose value is proportional to the phase difference between the two input signals. The phase difference between the two input signals is usually referred to as the error phase and is designated Θ_e.

Figure 2-9 shows a graph of dc output voltage V_{out} versus error phase Θ_e for a typical phase detector. Notice that the curve has a triangular shape with a negative slope from 0° to 180° (0 rad to π rad). As you

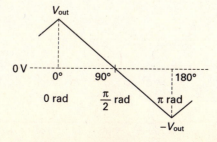

Fig. 2-9

can see, V_{out} has its maximum positive value when $\Theta_e = 0°$ and its maximum negative value when $\Theta_e = 180°$. At $\Theta_e = 90°$, $V_{out} = 0$ V dc. Recall that the VCO's natural frequency f_N occurs when its dc input voltage equals 0 V. Therefore, Θ_e must equal 90° when $f_i = f_N$. If f_i increases above f_N, then Θ_e increases and V_{out} becomes negative. Conversely, if f_i decreases below f_N, Θ_e decreases and V_{out} becomes positive. The dc voltage out of the phase detector is fed to a low-pass filter and dc amplifier before being applied as an input to the VCO. Nevertheless, the dc voltage out of the phase detector eventually reaches the input to the VCO.

Low-Pass Filter

Because a phase detector is a nonlinear circuit, its output is not a pure dc voltage. Instead, the input signals f_i and f_{VCO} mix and produce cross-product frequencies. One of the cross-product frequencies equals the sum of f_i and f_{VCO} ($f_i + f_{VCO}$) while another equals the difference between f_i and f_{VCO} ($f_i - f_{VCO}$). Therefore, at the output of the phase detector, the following frequencies exist: f_i, f_{VCO}, harmonics of f_i, and f_{VCO}, $f_i + f_{VCO}$, and $f_i - f_{VCO}$. It is the purpose of the low-pass filter to eliminate all frequencies except the difference frequency $f_i - f_{VCO}$. (The difference frequency $f_i - f_{VCO}$ is sometimes called the *beat frequency*.) The low-pass filter may be either a single-pole or a multiple-pole filter. As you will see in the discussions that follow, the frequency response of the low-pass filter plays a major role in the overall operation of the PLL.

DC Amplifier

The function of the dc amplifier is straightforward. It simply amplifies the filtered dc output voltage from the phase detector to a level that is appropriate for properly controlling the frequency of the VCO.

PLL Operation

Figure 2-10 shows the block diagram of a PLL without the dc amplifier. (Most block diagrams of a PLL do not include the dc amplifier.) The input voltage is represented as V_i and the input frequency as f_i. Similarly, V_o and f_o represent the VCO's output voltage and frequency respectively.

Assume that the input signal frequency f_i is initially equal to the natural frequency f_N of the VCO. If this is the case, then the dc input to the VCO must be 0 V, which in turn means that the error phase Θ_e between f_i and f_o must be 90° (refer to Figs. 2-8 and 2-9). If the input frequency f_i increases, then the error phase Θ_e between f_i and f_o also increases. The increased phase error causes the dc voltage at the phase-detector output to become negative. The negative output from the phase detector is filtered by the low-pass filter and then applied as an input to the VCO. With a negative dc voltage at the VCO input, the VCO frequency increases until $f_i = f_o$. The phase error Θ_e remains at a value greater than 90° as long as $f_i > f_N$. This is necessary since the dc input to the VCO must be negative if it is to oscillate at a frequency higher than its natural frequency f_N. Suppose now that the input signal frequency f_i decreases below the natural frequency f_N of the VCO. This causes the error phase Θ_e to decrease, thus producing a positive dc voltage at the phase-detector output. The positive dc voltage is filtered by the low-pass filter and applied

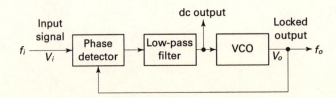

Fig. 2-10

as an input to the VCO. With a positive dc voltage at the VCO input, the output frequency decreases until $f_i = f_o$. The phase error Θ_e between f_i and f_o is maintained at some value less than 90° as long as $f_i < f_N$.

From the previous discussion, you should understand that a PLL locks the VCO frequency onto the input frequency. If the frequency of f_i changes, then the VCO frequency will track it. With f_o tracking f_i, the PLL is said to be in *phase lock*.

Lock Range

The range of frequencies over which a PLL can maintain phase lock is called its *lock range*. The lock-range width, designated B_L, can be specified as:

$$B_L = f_{\max} - f_{\min} \qquad (2.16)$$

The frequencies $f_{\max}$ and $f_{\min}$ represent the maximum and minimum frequencies over which phase lock can be maintained. The factors limiting the lock range include the maximum frequency deviation of the VCO and the dc voltage range of the phase-detector output. The lock range is independent of the low-pass filter's frequency response because, when the PLL is in phase lock, the difference frequency $f_i - f_o$ is zero.

Capture Range

The *capture range* width, B_C, refers to the range of frequencies over which a PLL can acquire phase lock when phase lock does not yet exist. If a PLL is not in phase lock, then $f_i \neq f_o$. For this condition $f_o = f_N$. Therefore, the capture-range width is a measure of how close to the natural frequency f_N the input-signal frequency f_i must come in order to obtain phase lock. Once phase lock has been acquired, the VCO output frequency f_o can track or follow the input frequency f_i over the entire lock range. The capture range is always within the lock range. The capture range of a PLL can be specified as:

$$B_C = f_2 - f_1 \qquad (2.17)$$

The frequency f_2 represents the highest frequency the PLL can lock onto, whereas f_1 represents the lowest frequency the PLL can lock onto. Figure 2-11 shows the relationship between the lock-range width, B_L, and capture-range width, B_C.

In order for a PLL to acquire phase lock, the difference frequency $f_i - f_N$ must not exceed the bandwidth of the low-pass filter. If it does, phase lock cannot be acquired. Figure 2-12 illustrates the idea. If the bandwidth of the low-pass filter is greater than the difference frequency $f_i - f_N$, then phase lock can be achieved. However, if the bandwidth is less than $f_i - f_N$, phase lock cannot be achieved. This is because the amplitude of the filter output voltage falls off sharply outside the bandwidth of the low-pass filter.

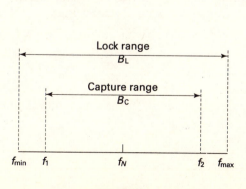

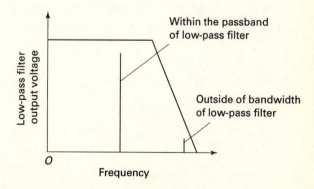

Fig. 2-11   Fig. 2-12

The NE565 PLL IC

Figure 2-13 shows the block diagram of an NE565, which is a 14-pin PLL integrated circuit (IC). The NE565 is designed for dual-supply operation and can operate with supply voltages from ± 5 V to ± 12 V. In Fig. 2-13 a positive voltage V_{CC} connects to Pin 10 and $-V_{CC}$ connects to Pin 1.

Pins 2 and 3 are differential inputs to the phase detector. Equal resistors with typical values ranging from 1 to 10 kΩ should be connected between each input and ground. These resistors provide a dc return path to ground and also determine the input impedance for the phase detector inputs. The input signal can be capacitively coupled to either Pin 2 or 3 of the phase detector. The capacitive reactance X_C of the input coupling capacitor should be one-tenth the input resistance at the lowest operating frequency. This can be expressed as:

$$C_{in} = \frac{10}{2\pi f_{min} R_{in}}$$
(2.18)

where C_{in} is the input coupling capacitance, f_{min} is the lowest operating frequency, and R_{in} is the dc input resistance between the phase-detector input and the ground.

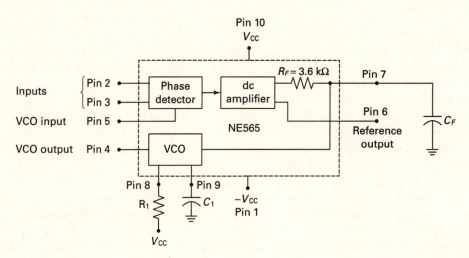

Fig. 2-13

For the NE565 PLL, the VCO's natural frequency f_N occurs when its dc input voltage equals $\frac{3}{4}V_{CC}$. The natural frequency, f_N is set by the timing resistor R_1 and the timing capacitor C_1. The timing resistor R_1 is connected between Pin 8 and the positive supply voltage V_{CC} whereas the timing capacitor C_1 is connected between Pin 9 and ground. With R_1 and C_1 known, f_N is calculated using the following equation.

$$f_N = \frac{1}{3.3 R_1 C_1}$$
(2.19)

The timing capacitor C_1 is normally chosen as a value between 100 pF and about 0.1 μF. With this in mind, Eq. (2.19) can be rearranged to solve for R_1:

$$R_1 = \frac{1}{3.3 f_N C_1}$$
(2.20)

The VCO output at Pin 4 is a square wave which varies from about 0 V to a peak value of approximately V_{CC}. A triangular voltage exists across the timing capacitor at Pin 9. The triangular voltage has a peak-to-peak amplitude of approximately one-third V_{CC}.

The VCO output from Pin 4 is usually connected to the VCO input at Pin 5. Therefore, the VCO output serves as the other input to the phase detector. Pin 4 also serves as the output from the PLL.

The internal 3.6 kΩ resistor R_F and an external capacitor C_F make up the PLL's low-pass filter. The purpose of the low-pass filter is to remove the original frequencies f_i and f_o, their harmonics, and the sum frequency $f_i + f_o$. The cutoff frequency f_c of the low-pass filter can be calculated as:

$$f_C = \frac{1}{2\pi R_F C_F} \tag{2.21}$$

The lower the cutoff frequency of the low-pass filter, the smaller the capture range. If the capacitor C_F is omitted, the capture range equals the lock range.

Pin 7, where the filter capacitor C_F is connected, is usually referred to as the FM output. This output is used only when a frequency modulated (FM) signal is driving the phase-detector input. The dc voltage available at Pin 7 equals $\frac{3}{4} V_{CC}$. For example, if $V_{CC} = 6$ V, then the dc voltage at Pin 7 equals $\frac{3}{4} \times 6$ V $= 4.5$ V. It is important to note, however, that the dc level of $\frac{3}{4} V_{CC}$ exists only when $f_i = f_N$. If $f_i < f_N$, then the dc level is greater than $\frac{3}{4} V_{CC}$. If $f_i > f_N$, then the dc level is less than $\frac{3}{4} V_{CC}$.

Pin 6, referred to as the reference output, provides a constant dc voltage of $\frac{3}{4} V_{CC}$. If Pins 6 and 7 drive an op-amp differencing amplifier, the dc level of $\frac{3}{4} V_{CC}$ appears as a common mode signal and is therefore eliminated. The result is a dc output of 0 V from the difference amplifier. When using the PLL as an FM demodulator, it is often desirable to eliminate the $\frac{3}{4} V_{CC}$ dc level.

The quantity B_L is the lock-range width which is the range of frequencies over which phase lock can be maintained. The lock range is centered on both sides of f_N, the VCO's natural frequency. B_L is calculated as:

$$B_L = \frac{16 f_N}{V_{CC(total)}} \tag{2.22}$$

In Eq. (2.22), $V_{CC(total)}$ represents the total supply voltage. For example, if $V_{CC} = 9$ V and $-V_{CC} = -9$ V, then the value of V_{CC} to use in Eq. (2.22) is 9 V $- (-9$ V$) = 18$ V. Equation (2.22) shows that the lock-range width B_L is inversely proportional to the total supply voltage V_{CC}. To achieve the full lock range the input voltage to the 565 should have an amplitude greater than 100 mV peak.

To calculate the capture-range width; use:

$$B_C = \sqrt{\frac{B_L}{\pi R_F C_F}} \tag{2.23}$$

Equation (2.23) can also be shown as:

$$B_C = \sqrt{2 B_L f_C} \tag{2.24}$$

where f_C represents the cutoff frequency of the low-pass filter.

Like B_L, B_C is centered on both sides of f_N. (Recall that, if the PLL is not in phase lock, then the VCO frequency equals the natural frequency f_N.)

EXAMPLE 2.7.   Refer to Fig. 2-14.

(a)   To what value must R_1 be adjusted to obtain a natural frequency f_N of 100 kHz?

(b)   With $f_N = 100$ kHz, calculate B_L and B_C.

(c)   Calculate the required value for the input coupling capacitor, C_{in}.

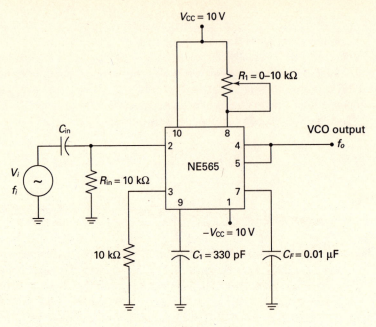

Fig. 2-14

Ans.

(*a*)   To calculate the value of R_1, use Eq. (*2.20*).

$$R_1 = \frac{1}{3.3 f_N C_1}$$

$$= \frac{1}{3.3 \times 100 \, \text{kHz} \times 330 \, \text{pF}}$$

$$= 9.18 \, \text{k}\Omega$$

(*b*)   Since $V_{CC} = \pm 10 \, \text{V}$, $V_{CC(\text{total})} = 20 \, \text{V}$. From Eq. (*2.22*), B_L is calculated as follows:

$$B_L = \frac{16 f_N}{V_{CC(\text{total})}}$$

$$= \frac{16 \times 100 \, \text{kHz}}{20 \, \text{V}}$$

$$= 80 \, \text{kHz}$$

or $$B_L = f_{\text{max}} - f_{\text{min}}$$
$$= 140 \, \text{kHz} - 60 \, \text{kHz}$$
$$= 80 \, \text{kHz}$$

To calculate the capture range B_C, use Eq. (*2.23*).

$$B_C = \sqrt{\frac{B_L}{\pi R_F C_F}}$$

$$= \sqrt{\frac{80 \, \text{kHz}}{\pi \times 3.6 \, \text{k}\Omega \times 0.01 \, \mu\text{F}}}$$

$$= 26.6 \, \text{kHz}$$

or $$B_C = f_2 - f_1$$
$$= 113.3 \, \text{kHz} - 86.7 \, \text{kHz}$$
$$= 26.6 \, \text{kHz}$$

(c)
$$C_{\text{in}} = \frac{10}{2\pi f_{\min} R_{\text{in}}}$$
$$= \frac{10}{2 \times \pi \times 60\,\text{kHz} \times 10\,\text{k}\Omega}$$
$$= 2.65\,\text{nF}$$

Note that $f_{\min}$ represents the lowest frequency the PLL can lock onto.

EXAMPLE 2.8. Calculate the capture range in Fig. 2-14 if C_F is changed to 0.02 μF.
   Ans.   Use Eq. *(2.23)*.

$$B_C = \sqrt{\frac{B_L}{\pi R_F C_F}}$$
$$= \sqrt{\frac{80\,\text{kHz}}{\pi \times 3.6\,\text{k}\Omega \times 0.02\,\mu\text{F}}}$$
$$= 18.8\,\text{kHz}$$
or
$$B_C = f_2 - f_1$$
$$= 109.4\,\text{kHz} - 90.6\,\text{kHz}$$
$$= 18.8\,\text{kHz}$$

Notice that lowering the cutoff frequency f_C reduces the capture-range width B_C.

2.4 FREQUENCY SYNTHESIZERS

A frequency synthesizer is essentially a frequency source whose output is an integer multiple of a very stable input reference frequency. A very basic frequency synthesizer, shown in Fig. 2-15, consists of a PLL with a divide-by-N counter inserted between the VCO output and phase detector input. The divide-by-N counter is a digital device that can be programmed with the desired binary number using thumbwheel switches. To make the PLL work correctly, f_i must equal f_o/N. This can be expressed as:

$$f_i = \frac{f_o}{N} \tag{2.25}$$

Therefore:
$$f_o = N f_i \tag{2.26}$$

By varying N, the output frequency f_o can be changed in increments of f_i. Therefore, if it is necessary to change f_o by increments of 10 kHz, the input reference frequency must equal 10 kHz. This frequency is usually generated by a highly stable crystal-controlled oscillator. In Fig 2-15, the phase detector, low-pass filter, and VCO make up the forward path of the loop while the "$\div N$" counter constitutes the feedback path.

In modern electronic communication systems, it is not uncommon for frequency synthesizers to produce frequencies in the 30 to 300 MHz range. Since the maximum operating frequency of most TTL and CMOS divide-by-N counters is about 50 MHz, the basic frequency synthesizer of Fig. 2-15 is not well

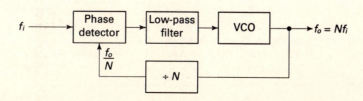

Fig. 2-15

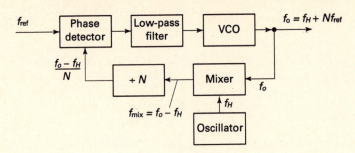

Fig. 2-16

suited for use at these very high frequencies. A technique often used to reduce the frequencies generated within the loop is shown in Fig. 2-16. This technique is commonly referred to as the *heterodyne-down conversion technique*. Notice that the VCO output frequency f_o and the local oscillator frequency f_H are fed into a mixer stage. The mixer is a nonlinear circuit which produces cross-product frequencies of f_o and f_H. Within the mixer, the following frequencies exist: f_o, f_H, harmonics of f_o and f_H, the sum frequency $f_o + f_H$, and the difference frequency $f_o - f_H$. The mixer output contains a filter which removes all frequencies except the difference frequency $f_o - f_H$. The difference frequency is applied to the input of the $\div N$ counter. To understand how the frequency synthesizer in Fig. 2-16 works, consider the following example.

EXAMPLE 2.9. Figure 2-17 shows a heterodyne-down conversion frequency synthesizer which produces one hundred different output frequencies in a range from 98.8 to 118.6 MHz. Each available output frequency is separated by 200 kHz. Calculate the minimum and maximum values for the $\div N$ factor.

Ans. First, the input reference frequency f_{ref} is 200 kHz which equals the spacing between each of the available output frequencies. Next, the VCO output frequency f_o and the oscillator output f_H are fed into a mixer. The output of the mixer equals the difference between f_o and f_H, or $f_{mix} = f_o - f_H$. The difference frequency is then fed into the $\div N$ counter. The minimum and maximum values for the $\div N$ factor are calculated as:

$$N_{(min)} = \frac{f_{o(min)} - f_H}{f_{ref}}$$

$$= \frac{98.8\,\text{MHz} - 98\,\text{MHz}}{200\,\text{kHz}}$$

$$= \frac{800\,\text{kHz}}{200\,\text{kHz}}$$

$$N_{(min)} = 4$$

$$N_{(max)} = \frac{f_{o(max)} - f_H}{f_{ref}}$$

$$= \frac{118.6\,\text{MHz} - 98\,\text{MHz}}{200\,\text{kHz}}$$

$$= \frac{20.6\,\text{MHz}}{200\,\text{kHz}}$$

$$N_{(max)} = 103$$

These calculations indicate that the divide-by-N counter must be capable of counting by any integer number between 4 and 103. In fact, to tune from one frequency to the next, the $\div N$ factor must be changed to the appropriate value. For example, if the $\div N$ counter is set to divide by 60, then f_o equals:

$$f_o = Nf_{ref} + f_H$$

$$= (60 \times 200\,\text{kHz}) + 98\,\text{MHz}$$

$$= 12\,\text{MHz} + 98\,\text{MHz}$$

$$= 110\,\text{MHz}$$

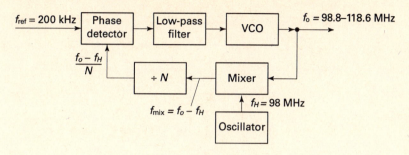

Fig. 2-17

This value was obtained by rearranging the following equation and solving for f_o.

$$N = \frac{f_o - f_H}{f_{ref}}$$

In some cases, the reference oscillator frequency may need to be reduced in value prior to entering the phase detector. In this case, the reference oscillator is also fed through a divide-by-N counter as shown in Fig. 2-18.

Figure 2-19 shows a frequency synthesizer which uses a technique known as prescaling. With this technique, a fixed-modulus counter prescales or reduces the VCO output frequency f_o by a constant factor k, to a value that can easily be handled by the integrated circuit(s) used for the programmable counter. The fixed-modulus prescaled counter is made up of very fast logic elements such as emitter-coupled logic (ECL) units. In Fig. 2-19, notice that the phase detector has input frequencies of f_{ref} and f_o/KN. Therefore:

$$f_{ref} = \frac{f_o}{KN} \tag{2.27}$$

With this technique, the output frequencies from the synthesizer are spaced by an amount equal to Kf_{ref} or:

$$f_{ch} = Kf_{ref} \tag{2.28}$$

where f_{ch} represents the increments by which f_o changes as N changes. With this being the case, the minimum and maximum values for the $\div N$ factor are calculated as follows:

$$N_{(min)} = \frac{f_{o(min)}}{Kf_{ref}} = \frac{f_{o(min)}}{f_{ch}}$$

$$N_{(max)} = \frac{f_{o(max)}}{Kf_{ref}} = \frac{f_{o(max)}}{f_{ch}}$$

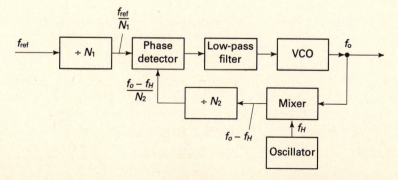

Fig. 2-18

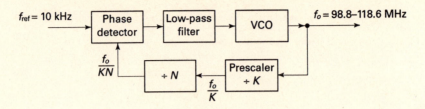

Fig. 2-19

To understand the operation of a frequency synthesizer which uses prescaling, see the following example.

EXAMPLE 2.10.   Figure 2-19 shows a prescaling frequency synthesizer which produces one hundred different output frequencies in a range from 98.8 to 118.6 MHz. Each frequency is to be separated by 200 kHz. Determine the required divide-by-K factor as well as the minimum and maximum values for the $\div N$ counter. The reference frequency f_{ref} equals 10 kHz.

Ans.   First, determine the fixed value of K by rearranging Eq. (*2-28*). Begin with:

$$f_{ch} = K f_{ref}$$
$$K = \frac{f_{ch}}{f_{ref}}$$
$$= \frac{200\,\text{kHz}}{10\,\text{kHz}}$$
$$= 20$$

Next, calculate $N_{(min)}$ and $N_{(max)}$.

$$N_{(min)} = \frac{f_{o(min)}}{f_{ch}}$$
$$= \frac{98.8\,\text{MHz}}{200\,\text{kHz}}$$
$$= 494$$
$$N_{(max)} = \frac{f_{o(max)}}{f_{ch}}$$
$$= \frac{118.6\,\text{MHz}}{200\,\text{kHz}}$$
$$= 593$$

The advantage of a prescaling frequency synthesizer versus the heterodyne-down conversion type is that there is no need for an additional oscillator and mixer to achieve the desired output frequencies.

Solved Problems

2.1   Refer to the block diagram in Fig. 2-20. If $A_V = 10$ and $B = 0.02$, will the circuit oscillate? If not, calculate the required value of A_V.

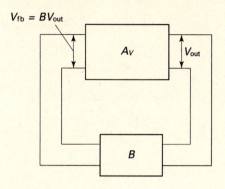

$V_{fb} = BV_{out}$

A_V

V_{out}

B

Fig. 2-20

SOLUTION

Calculate the $A_V B$ product.

$$A_V B = 10 \times 0.02$$
$$= 0.2$$

Since $A_V B < 1$, the Barkhausen criterion is not satisfied, and therefore oscillations will not exist. In order for oscillations to occur, the following must be true.

$$A_V \geq \frac{1}{B}$$
$$= \frac{1}{0.02}$$
$$= 50$$

2.2   Refer to Fig. 2-20. If $A_V = 150$ and $B = 0.01$, will the circuit oscillate?

SOLUTION

Determine if $A_V B \geq 1$.

$$A_V B = 150 \times 0.01$$
$$= 1.5$$

Since $A_V B > 1$, the circuit will oscillate.

2.3   Refer to Fig. 2-20. Assume $B = -0.025$. Calculate the value of A_V required to satisfy the Barkhausen criterion.

SOLUTION

Recall that the Barkhausen criterion states that the $A_V B$ product must be equal to 1. Solving for A_V, we have:

$$A_V = \frac{1}{B}$$
$$= \frac{1}{-0.025}$$
$$= -40$$

2.4 Assume in Fig. 2-20 that $V_{out} = 24$ V p–p (volts peak-to-peak) and $B = 0.01$. Calculate A_V and V_{in}.

SOLUTION

Begin by calculating V_{in}.

$$V_{in} = V_{fb} = BV_{out}$$
$$= 0.01 \times 24 \text{ V p–p}$$
$$= 240 \text{ mV p–p}$$

Next,
$$A_V = \frac{V_{out}}{V_{in}}$$
$$= \frac{24 \text{ V p–p}}{240 \text{ mV p–p}}$$
$$= 100$$

2.5 Refer to Fig. 2-21. Assume $L_1 = 496.67$ μH and $L_2 = 9.93$ μH. Calculate the following: (a) f_{osc}; (b) the feedback fraction B; (c) X_{L_1}, X_{L_2}, and X_{C_3}; (d) X_T (net reactance) of the $L_2 C_3$ branch.

SOLUTION

(a) Begin by calculating the total tank inductance L_T.

$$L_T = L_1 + L_2$$
$$= 496.67 \text{ }\mu\text{H} + 9.93 \text{ }\mu\text{H}$$
$$= 506.6 \text{ }\mu\text{H}$$

Therefore,
$$f_{osc} = \frac{1}{2\pi\sqrt{L_T C_3}}$$
$$= \frac{1}{2\pi\sqrt{506.6 \text{ }\mu\text{H} \times 50 \text{ pF}}}$$
$$= 1 \text{ MHz}$$

(b)
$$B = \frac{L_2}{L_1}$$
$$= \frac{9.93 \text{ }\mu\text{H}}{496.67 \text{ }\mu\text{H}}$$
$$= 0.02 \text{ or } \tfrac{1}{50}$$

(c)
$$X_{L_1} = 2\pi f_{osc} L_1$$
$$= 2 \times \pi \times 1 \text{ MHz} \times 496.67 \text{ }\mu H$$
$$= 3.12 \text{ k}\Omega$$
$$X_{L_2} = 2\pi f_{osc} L_2$$
$$= 2 \times \pi \times 1 \text{ MHz} \times 9.93 \text{ }\mu H$$
$$= 62.4 \text{ }\Omega$$
$$X_{C_3} = \frac{1}{2\pi f_{osc} C_3}$$
$$= \frac{1}{2 \times \pi \times 1 \text{ MHz} \times 50 \text{ pF}}$$
$$= 3.18 \text{ k}\Omega$$

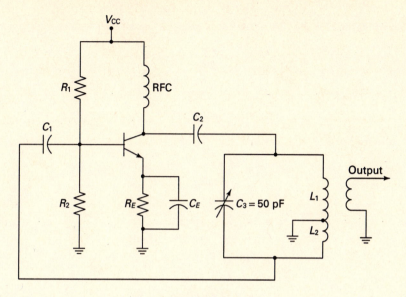

Fig. 2-21

(d)
$$X_T = X_{C_3} - X_{L_2}$$
$$= 3.18\,\text{k}\Omega - 62.4\,\Omega$$
$$= 3.12\,\text{k}\Omega$$

Notice how
$$X_{L_1} = X_{C_3} - X_{L_2}$$
$$= 3.18\,\text{k}\Omega - 62.4\,\Omega$$
$$= 3.12\,\text{k}\Omega$$

2.6 Refer to Fig. 2-21. Assume that it is desired to obtain an oscillating frequency of 500 kHz. If $B = \frac{1}{25}$, calculate: (a) L_T; (b) L_1 and L_2; (c) X_{L_1}, X_{L_2}, and X_{C_3}; (d) X_T of the $L_2 C_3$ branch.

SOLUTION

(a)
$$L_T = \frac{1}{4\pi^2 f_{\text{osc}}^2 C_3}$$
$$= \frac{1}{4 \times \pi^2 \times 500^2\,\text{kHz} \times 50\,\text{pF}}$$
$$= 2.03\,\text{mH}$$

(b) Since $B = L_2/L_1$ and $B = \frac{1}{25}$:
$$\frac{1}{25} = \frac{L_2}{L_1}$$

Rearranging to solve for L_1 gives us:
$$L_1 = 25L_2$$

Next, write the expression for L_T.
$$L_T = L_1 + L_2$$

Next, substitute $25L_2$ for L_1:
$$L_T = 25L_2 + L_2$$

which simplifies to $L_T = 26L_2$. Next, solve for L_2.

$$L_2 = \frac{L_T}{26}$$

$$= \frac{2.03 \text{ mH}}{26}$$

$$= 78.1 \text{ } \mu\text{H}$$

Next, solve for L_1.

$$L_1 = L_T - L_2$$

$$= 2.03 \text{ mH} - 78.1 \text{ } \mu\text{H}$$

$$= 1.95 \text{ mH}$$

(c) $$X_{L_1} = 2\pi f_{\text{osc}} L_1$$

$$= 2 \times \pi \times 500 \text{ kHz} \times 1.95 \text{ mH}$$

$$= 6.13 \text{ k}\Omega$$

$$X_{L_2} = 2\pi f_{\text{osc}} L_2$$

$$= 2 \times \pi \times 500 \text{ kHz} \times 78.1 \text{ } \mu\text{H}$$

$$= 245.4 \text{ }\Omega$$

$$X_{C_3} = \frac{1}{2 \times \pi \times 500 \text{ kHz} \times 50 \text{ pF}}$$

$$= 6.37 \text{ k}\Omega$$

(d) $$X_T = X_{C_3} - X_{L_2}$$

$$= 6.37 \text{ k}\Omega - 245.4 \text{ }\Omega$$

$$= 6.13 \text{ k}\Omega$$

Notice that X_{L_1} equals the net reactance (X_T) of the $L_2 C_3$ branch.

2.7   Refer to Fig. 2-22. If $C_1 = 14.54$ pF and $C_2 = 435.27$ pF, calculate the following: (a) f_{osc}; (b) the feedback fraction B; (c) X_{C_1}, X_{C_2}, and X_{L_1}; (d) X_T of the $L_1 C_2$ branch.

SOLUTION

(a)   Begin by calculating C_{eq}.

$$C_{\text{eq}} = \frac{C_1 C_2}{C_1 + C_2}$$

$$= \frac{14.54 \text{ pF} \times 435.27 \text{ pF}}{14.54 \text{ pF} + 435.27 \text{ pF}}$$

$$= 14.07 \text{ pF}$$

Next, $$f_{\text{osc}} = \frac{1}{2\pi\sqrt{LC_{\text{eq}}}}$$

$$= \frac{1}{2 \times \pi\sqrt{200 \text{ } \mu\text{H} \times 14.07 \text{ pF}}}$$

$$= 3 \text{ MHz}$$

(b) $$B = \frac{C_1}{C_2}$$

$$= \frac{14.54 \text{ pF}}{435.27 \text{ pF}}$$

$$= 0.0334 \text{ or } \tfrac{1}{30}$$

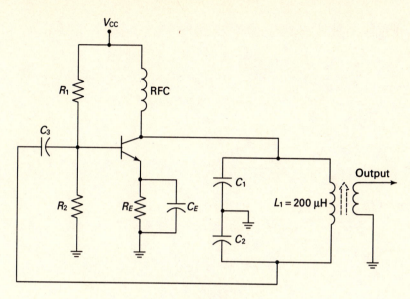

Fig. 2-22

(c)
$$X_{C_1} = \frac{1}{2\pi f_{osc} C_1}$$

$$= \frac{1}{2 \times \pi \times 3\,\text{MHz} \times 14.54\,\text{pF}}$$

$$= 3.65\,\text{k}\Omega$$

$$X_{C_2} = \frac{1}{2\pi f_{osc} C_2}$$

$$= \frac{1}{2 \times \pi \times 3\,\text{MHz} \times 435.27\,\text{pF}}$$

$$= 121.88\,\Omega$$

$$X_{L_1} = 2\pi f_{osc} L_1$$

$$= 2 \times \pi \times 3\,\text{MHz} \times 200\,\mu\text{H}$$

$$= 3.77\,\text{k}\Omega$$

(d)
$$X_T = X_{L_1} - X_{C_2}$$

$$= 3.77\,\text{k}\Omega - 121.88\,\Omega$$

$$= 3.65\,\text{k}\Omega$$

2.8   Refer to Fig. 2-22. Assume that it is desired to obtain an oscillating frequency of 1 MHz and a feedback fraction B of $\frac{1}{25}$. Calculate the following: (a) C_{eq}; (b) C_1 and C_2; (c) X_{C_1}, X_{C_2}, and X_{L_1}; (d) X_T of the $L_1 C_2$ branch.

SOLUTION

(a)
$$C_{eq} = \frac{1}{4\pi^2 f_{osc}^2 L}$$

$$= \frac{1}{4 \times \pi^2 \times 1^2\,\text{MHz} \times 200\,\mu\text{H}}$$

$$= 126.65\,\text{pF}$$

(b) Recall that, for a Colpitts oscillator, $B = C_1/C_2$. Therefore:

$$\frac{1}{25} = \frac{C_1}{C_2}$$

Solving for C_2 gives us:

$$C_2 = 25C_1$$

Next,

$$\frac{1}{C_{eq}} = \frac{1}{C_1} + \frac{1}{25C_1}$$

$$\frac{1}{C_{eq}} = \frac{26}{25C_1}$$

Solving for C_1 gives us:

$$C_1 = C_{eq} \times \frac{26}{25}$$

$$= 126.65\,\text{pF} \times \frac{26}{25}$$

$$= 131.72\,\text{pF}$$

And finally,

$$C_2 = \frac{1}{\dfrac{1}{C_{eq}} - \dfrac{1}{C_1}}$$

$$= \frac{1}{\dfrac{1}{126.65\,\text{pF}} - \dfrac{1}{131.72\,\text{pF}}}$$

$$= 3.29\,\text{nF}$$

(c)

$$X_{C_1} = \frac{1}{2\pi f_{osc} C_1}$$

$$= \frac{1}{2 \times \pi \times 1\,\text{MHz} \times 131.72\,\text{pF}}$$

$$= 1.208\,\text{k}\Omega$$

$$X_{C_2} = \frac{1}{2\pi f_{osc} C_2}$$

$$= \frac{1}{2 \times \pi \times 1\,\text{MHz} \times 3.29\,\text{nF}}$$

$$= 48.38\,\Omega$$

$$X_{L_1} = 2\pi f_{osc} L_1$$

$$= 2 \times \pi \times 1\,\text{MHz} \times 200\,\mu\text{H}$$

$$= 1.256\,\text{k}\Omega$$

(d)

$$X_T = X_{L_1} - X_{C_2}$$

$$= 1.256\,\text{k}\Omega - 48.38\,\Omega$$

$$= 1.208\,\text{k}\Omega$$

2.9 In a Colpitts oscillator, C_1 and C_2 are shunted by transistor and stray capacitances. This additional capacitance slightly alters the frequency of oscillation. Figure 2-23 shows how the oscillating frequency can be made nearly independent of the transistor and stray capacitances. An additional capacitor C_3 has been added to the tank circuit. This capacitor is much smaller than either C_1 or C_2. As a result, the

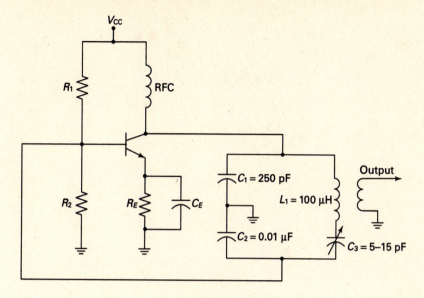

Fig. 2-23

equivalent tank capacitance is approximately equal to C_3. This type of oscillator is known as a *Clapp oscillator*. If C_3 is adjusted to 10.57 pF, calculate the following: (*a*) the feedback fraction B; (*b*) f_{osc}; (*c*) X_{C_1}, X_{C_2}, X_{C_3}, and X_{L_1}; (*d*) X_T of the $L_1 C_2 C_3$ branch.

SOLUTION

(*a*) The feedback fraction B is not affected by the addition of C_3. Therefore,

$$B = \frac{C_1}{C_2}$$

$$= \frac{250\,\text{pF}}{0.01\,\mu\text{F}}$$

$$= 0.025 \text{ or } \tfrac{1}{40}$$

(*b*) To calculate f_{osc}, first determine C_{eq}.

$$C_{eq} = \frac{1}{\dfrac{1}{C_1} + \dfrac{1}{C_2} + \dfrac{1}{C_3}}$$

$$= \frac{1}{\dfrac{1}{250\,\text{pF}} + \dfrac{1}{0.01\,\mu\text{F}} + \dfrac{1}{10.57\,\text{pF}}}$$

$$= 10.13\,\text{pF}$$

Next,

$$f_{osc} = \frac{1}{2\pi\sqrt{L_1 C_{eq}}}$$

$$= \frac{1}{2 \times \pi\sqrt{100\,\mu\text{H} \times 10.13\,\text{pF}}}$$

$$= 5\,\text{MHz}$$

(c)
$$X_{C_1} = \frac{1}{2\pi f_{osc} C_1}$$

$$= \frac{1}{2 \times \pi \times 5\,\text{MHz} \times 250\,\text{pF}}$$

$$= 127.32\,\Omega$$

$$X_{C_2} = \frac{1}{2\pi f_{osc} C_2}$$

$$= \frac{1}{2 \times \pi \times 5\,\text{MHz} \times 0.01\,\mu\text{F}}$$

$$= 3.18\,\Omega$$

$$X_{C_3} = \frac{1}{2\pi f_{osc} C_3}$$

$$= \frac{1}{2 \times \pi \times 5\,\text{MHz} \times 10.57\,\text{pF}}$$

$$= 3.01\,\text{k}\Omega$$

$$X_{L_1} = 2\pi f_{osc} L_1$$

$$= 2 \times \pi \times 5\,\text{MHz} \times 100\,\mu\text{H}$$

$$= 3.14\,\text{k}\Omega$$

(d)
$$X_T = X_{L_1} - X_{C_2} - X_{C_3}$$

$$= 3.14\,\text{k}\Omega - 3.18\,\Omega - 3.01\,\text{k}\Omega$$

$$= 126.82\,\Omega$$

Note that $X_{C_1} \cong X_T$. The small difference is due to rounding.

The Clapp oscillator is more stable than a Colpitts oscillator because the oscillating frequency is affected very little by the transistor and stray capacitance.

2.10 Refer to Fig. 2-24. Calculate the following: (a) f_{osc}, (b) the feedback fraction B, (c) the minimum value of A_V required to sustain oscillations.

SOLUTION

(a) To calculate the oscillating frequency, the effect of C_1 and C_2 can be ignored since their combined equivalent capacitance $C_1 C_2/(C_1 + C_2)$ is more than 100 times greater than that of C_3. Therefore:

$$f_{osc} = \frac{1}{2\pi\sqrt{L_1 C_3}}$$

$$= \frac{1}{2 \times \pi\sqrt{50\,\mu\text{H} \times 9\,\text{pF}}}$$

$$= 7.5\,\text{MHz}$$

(b)
$$B = \frac{C_1}{C_2}$$

$$= \frac{0.001\,\mu\text{F}}{0.1\,\mu\text{F}}$$

$$= 0.01 \text{ or } \tfrac{1}{100}$$

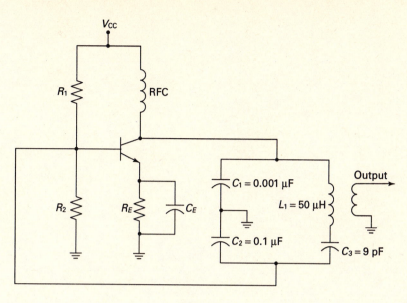

Fig. 2-24

(c)   Since $A_V B$ must equal 1,

$$A_V = \frac{1}{B}$$

$$= \frac{1}{0.01}$$

$$= 100$$

Actually, since a common-emitter amplifier is used in Fig. 2-24, both A_V and B must be negative quantities. This means the amplifier and feedback network must provide a total phase shift of 360°.

2.11   Refer to Fig. 2-25. In which mode does the crystal Y_1 operate? Does the resonant frequency of the tank affect circuit operation?

SOLUTION

The crystal Y_1 operates in its series resonant mode. Therefore, the frequency of oscillation occurs at the series resonant frequency f_s of the crystal. At f_s the crystal has a very low impedance, thus allowing the maximum amount of feedback to sustain oscillations.

The LC tank circuit is tuned to the series resonant frequency of the crystal. This is necessary to provide the circuit with the required voltage gain. At resonance, the tank circuit has maximum impedance. If the tank is tuned to a frequency too far above or below f_s, A_V will decrease to a value too low to sustain oscillations. Recall that for an unswamped common-emitter amplifier, $A_V = -Z_C/r_e'$, where Z_C represents the ac collector impedance, and r_e' represents the ac resistance of the base-emitter diode.

It should also be noted that L_2 and C_3 still provide the required 180° phase shift required of the feedback network. The crystal will not introduce any additional phase shift, since it appears purely resistive when operating in its series resonant mode.

2.12   Refer to Fig. 2-25. If Y_1 has a series resonant frequency of 1 MHz, calculate the following: (a) the value of C_2 required to resonate at 1 MHz, (b) the feedback fraction B, (c) the minimum voltage gain $A_{V(min)}$.

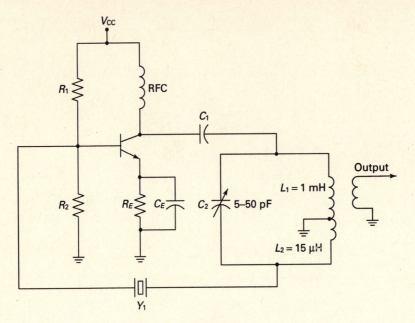

Fig. 2-25

SOLUTION

(a)

$$C_2 = \frac{1}{4\pi^2 f_{osc}^2 (L_1 + L_2)}$$

$$= \frac{1}{4 \times \pi^2 \times 1^2 \, \text{MHz} \times (1 \, \text{mH} + 15 \, \mu\text{H})}$$

$$= 24.96 \, \text{pF}$$

(b)

$$B = \frac{L_2}{L_1}$$

$$= \frac{15 \, \mu\text{H}}{1 \, \text{mH}}$$

$$= 0.015 \, \text{or} \, \frac{1}{66.67}$$

(c)

$$A_{V(\text{min})} = \frac{1}{B}$$

$$= \frac{1}{0.015}$$

$$= 66.67$$

2.13   Explain the operation of the crystal oscillator in Fig. 2-26.

SOLUTION

The operation of the crystal oscillator in Fig. 2-26 is similar to the operation of the Colpitts oscillator discussed earlier. The crystal Y_1 operates between f_s and f_p and thus acts like an inductance. As a result, Y_1 and C_2 provide the 180° phase shift required of the feedback network. C_1 and C_2 still determine the feedback fraction B. The stability is exceptional since oscillations will cease if the frequency should shift too close to either f_s or f_p.

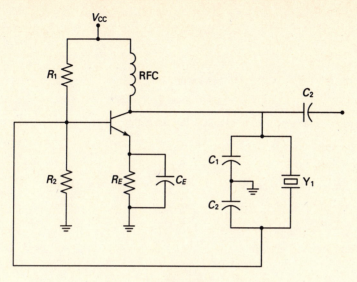

Fig. 2-26

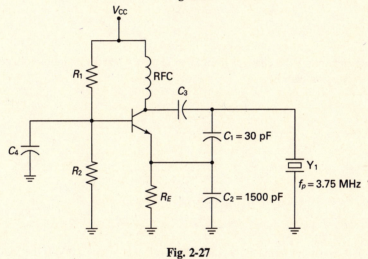

Fig. 2-27

2.14   Refer to Fig. 2-27. Calculate the feedback fraction B.

SOLUTION

Recall that the feedback fraction B for this circuit is calculated as:

$$B = \frac{C_{eq}}{C_2}$$

C_{eq} is calculated as:

$$C_{eq} = \frac{C_1 C_2}{C_1 + C_2}$$

$$= \frac{30 \, \text{pF} \times 1500 \, \text{pF}}{30 \, \text{pF} + 1500 \, \text{pF}}$$

$$= 29.41 \, \text{pF}$$

Therefore,

$$B = \frac{C_{eq}}{C_2}$$

$$= \frac{29.41 \text{ pF}}{1500 \text{ pF}}$$

$$= 0.0196 \text{ or } \tfrac{1}{51}$$

Another equation can also be used to calculate B. Start with:

$$B = \frac{C_{eq}}{C_2}$$

Expand C_{eq} to $C_1 C_2/C_1 + C_2$ and simplify:

$$B = \frac{C_1}{C_1 + C_2}$$

Inserting values from Fig. 2-27 gives us:

$$B = \frac{30 \text{ pF}}{30 \text{ pF} + 1500 \text{ pF}}$$

$$= 0.0196 \text{ or } \tfrac{1}{51}$$

2.15 The output frequency of a VCO changes from 50 to 75 kHz when the dc input voltage changes by 0.5 V. Calculate the conversion gain K.

SOLUTION

$$K = \frac{\Delta f}{\Delta V}$$

$$= \frac{75 \text{ kHz} - 50 \text{ kHz}}{0.5 \text{ V}}$$

$$= 50 \text{ kHz/V}$$

2.16 A certain VCO has a natural frequency f_N of 1 MHz with a dc input of 0 V. If the VCO has a conversion gain of 50 kHz/V, calculate the VCO output frequency f_o for a voltage change of 1 V.

SOLUTION

Rearrange the formula $K = \Delta f/\Delta V$:

$$\Delta f = K \Delta V$$

$$= 50 \text{ kHz/V} \times 1 \text{ V}$$

$$\Delta f = 50 \text{ kHz}$$

The VCO frequency equals:

$$f_o = f_N \pm \Delta f$$

$$= 1 \text{ MHz} \pm 50 \text{ kHz}$$

Whether f_o is at 950 kHz or 1.05 MHz depends on the transfer-function slope.

2.17 Refer to Fig. 2-28. If R_B is adjusted to 262.25 Ω, calculate the following: (a) f_N, (b) the cutoff frequency f_c of the low-pass filter, (c) the lock-range width B_L, (d) the capture-range width B_C, (e) the minimum value for C_1.

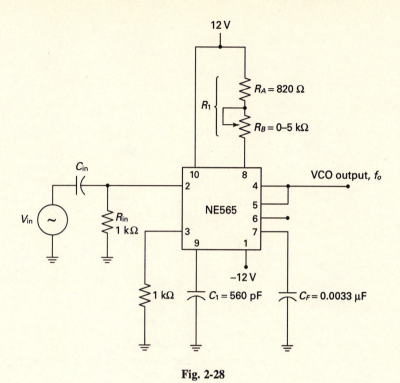

Fig. 2-28

SOLUTION

(*a*)   Begin by calculating R_1.

$$R_1 = R_A + R_B$$
$$= 820\,\Omega + 262.25\,\Omega$$
$$= 1.082\,\text{k}\Omega$$

Next, calculate f_N.

$$f_N = \frac{1}{3.3R_1C_1}$$
$$= \frac{1}{3.3 \times 1.082\,\text{k}\Omega \times 560\,\text{pF}}$$
$$= 500\,\text{kHz}$$

(*b*)

$$f_C = \frac{1}{2\pi R_F C_F}$$
$$= \frac{1}{2 \times \pi \times 3.6\,\text{k}\Omega \times 0.0033\,\mu\text{F}}$$
$$= 13.4\,\text{kHz}$$

(*c*)   First, $V_{\text{CC(total)}} = 12\,\text{V} - (-12\,\text{V}) = 24\,\text{V}$. Next,

$$B_L = \frac{16f_N}{V_{\text{CC(total)}}}$$
$$= \frac{16 \times 500\,\text{kHz}}{24\,\text{V}}$$
$$= 333.33\,\text{kHz}$$

or

$$B_L = f_{max} - f_{min}$$
$$= 666.67\,\text{kHz} - 333.33\,\text{kHz}$$
$$= 333.33\,\text{kHz}$$

(d)

$$B_C = \sqrt{\frac{B_L}{\pi R_F C_F}}$$

$$= \sqrt{\frac{333.33\,\text{kHz}}{\pi \times 3.6\,\text{k}\Omega \times 0.0033\,\mu\text{F}}}$$

$$= 94.5\,\text{kHz}$$

or

$$B_C = f_2 - f_1$$
$$= 547.25\,\text{kHz} - 452.75\,\text{kHz}$$
$$= 94.5\,\text{kHz}$$

(e)

$$C_{in} = \frac{10}{2\pi f_{min} R_{in}}$$

Since the lowest frequency the PLL can lock onto is 333.33 kHz,

$$C_{in} = \frac{10}{2 \times \pi \times 333.33\,\text{kHz} \times 1\,\text{k}\Omega}$$

$$= 4.78\,\text{nF}$$

2.18 In Fig. 2-28, assume that C_F is changed to 1000 pF. Recalculate the capture range B_C. (Recall that the lock-range width B_L was calculated to be 333.33 kHz.)

SOLUTION

$$B_C = \sqrt{\frac{B_L}{\pi R_F C_F}}$$

$$= \sqrt{\frac{333.33\,\text{kHz}}{\pi \times 3.6\,\text{k}\Omega \times 1000\,\text{pF}}}$$

$$= 171.68\,\text{kHz}$$

or

$$B_C = f_2 - f_1$$
$$= 585.84\,\text{kHz} - 414.16\,\text{kHz}$$
$$= 171.68.\,\text{kHz}$$

Notice that extending the cutoff frequency of the low-pass filter increases the capture range.

2.19 Refer to Fig. 2-28. To what value must the resistance R_B be adjusted to provide a natural frequency f_N of 100 kHz? What is the lock-range width B_L for $f_N = 100$ kHz?

SOLUTION

To calculate the required value of R_1, rearrange the equation for f_N.

$$R_1 = \frac{1}{3.3 f_N C_1}$$

$$= \frac{1}{3.3 \times 100\,\text{kHz} \times 560\,\text{pF}}$$

$$= 5.41\,\text{k}\Omega$$

Since $R_1 = R_A + R_B$, R_B must be calculated as follows:

$$R_B = R_1 - R_A$$
$$= 5.41 \, k\Omega - 820 \, \Omega$$
$$= 4.59 \, k\Omega$$

The lock-range width B_L equals:

$$B_L = \frac{16 f_N}{V_{CC(total)}}$$
$$= \frac{16 \times 100 \, kHz}{24 \, V}$$
$$= 66.67 \, kHz$$

2.20 In Fig. 2-28, assume $f_i = f_N$. Calculate the dc voltage available at Pin 6.

SOLUTION

The dc voltage available at Pin 6 equals $\frac{3}{4} V_{CC}$ which is calculated as:

$$V_{dc(Pin\,6)} = \frac{3}{4} \times V_{CC}$$
$$= \frac{3}{4} \times 12 \, V$$
$$= 9 \, V$$

2.21 The frequency synthesizer shown in Fig. 2-29 provides 400 output frequencies equally spaced by 10 kHz. The output frequencies extend from 144.0 to 148.0 MHz. The input reference frequency f_{ref} equals 10 kHz, and the high-frequency oscillator has a frequency f_H equal to 100 MHz. Calculate the minimum and maximum values required of the $\div N$ counter.

SOLUTION

Recall that $f_{ref} = (f_o - f_H)/N$. Solving for N gives us:

$$N = \frac{f_o - f_H}{f_{ref}}$$

To calculate N_{min} and N_{max}, proceed as follows.

$$N_{min} = \frac{f_{o(min)} - f_H}{f_{ref}}$$
$$= \frac{144 \, MHz - 100 \, MHz}{10 \, kHz}$$
$$= 4400$$

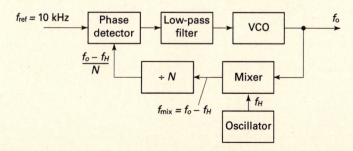

Fig. 2-29

$$N_{max} = \frac{f_{o(max)} - f_H}{f_{ref}}$$

$$= \frac{148\,\text{MHz} - 100\,\text{MHz}}{10\,\text{kHz}}$$

$$= 4800$$

2.22 Referring to Problem 2.21, what is the output frequency if $N = 4609$?

SOLUTION

Recall that $f_o = f_H + Nf_{ref}$. Therefore:

$$f_o = 100\,\text{MHz} + (4609 \times 10\,\text{kHz})$$

$$= 100\,\text{MHz} + 46.09\,\text{MHz}$$

$$= 146.09\,\text{MHz}$$

2.23 Referring to Problem 2.21, what changes need to be made if it is desired to produce 800 frequencies equally spaced by 5 kHz?

SOLUTION

Since the reference frequency f_{ref} determines how far apart the output frequencies are spaced, f_{ref} must be changed to 5 kHz. Next, calculate the minimum and maximum values for the $\div N$ factor.

$$N_{min} = \frac{f_{o(min)} - f_H}{f_{ref}}$$

$$= \frac{144\,\text{MHz} - 100\,\text{MHz}}{5\,\text{kHz}}$$

$$= 8800$$

$$N_{max} = \frac{f_{o(max)} - f_H}{f_{ref}}$$

$$= \frac{148\,\text{MHz} - 100\,\text{MHz}}{5\,\text{kHz}}$$

$$= 9600$$

2.24 The prescaling frequency synthesizer of Fig. 2-30 produces 20 output frequencies in the frequency range extending from 146.0 to 148.0 MHz. The frequencies are equally spaced by 100 kHz. Determine: (*a*) the required divide-by-K factor, (*b*) N_{min} and N_{max}.

SOLUTION

(*a*) Since $f_{ref} = 10\,\text{kHz}$ and $f_{ch} = 100\,\text{kHz}$, with $f_{ch} = Kf_{ref}$, we have:

$$K = \frac{f_{ch}}{f_{ref}}$$

$$= \frac{100\,\text{kHz}}{10\,\text{kHz}}$$

$$= 10$$

(*b*) Recall that $f_{ref} = f_o/KN$. Solving for N gives us:

$$N = \frac{f_o}{Kf_{ref}}$$

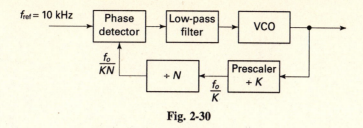

Fig. 2-30

To solve for $N_{\min}$ and $N_{\max}$, use the minimum and maximum values for f_o in the equation $N = f_o/Kf_{\text{ref}}$.

$$N_{\min} = \frac{f_{o(\min)}}{Kf_{\text{ref}}}$$

$$= \frac{146.0\,\text{MHz}}{10 \times 10\,\text{kHz}}$$

$$= 1460$$

$$N_{\max} = \frac{f_{o(\max)}}{Kf_{\text{ref}}}$$

$$= \frac{148.0\,\text{MHz}}{10 \times 10\,\text{kHz}}$$

$$= 1480$$

Supplementary Problems

2.25 Refer to the block diagram in Fig. 2-20. If $A_V = 50$ and $B = 0.01$, will the circuit oscillate? If not, calculate the required value of A_V.
Ans.   No, the circuit will not oscillate ($A_V B < 1$). A_V must be equal to or greater than 100 to sustain oscillations.

2.26 Refer again to Fig. 2-20. If $A_V = 50$ and $B = 0.04$, will the circuit oscillate?   *Ans.*   Yes ($A_V B = 2$)

2.27 Refer again to Fig. 2-20. If $B = -0.05$, what value of A_V is required to sustain oscillations?
Ans. $A_{V(\min)} = -20$

2.28 Refer to Fig. 2-20. Assume $A_V = 100$ and $B = 0.01$. Calculate v_{fb} if $V_{\text{out}} = 30$ V p-p.   *Ans.* $v_{\text{fb}} = 0.3$ V p–p

2.29 Refer to Fig. 2-31. Calculate the following: (a) f_{osc}; (b) the feedback fraction B; (c) X_{L_1}, X_{L_2}, and X_{C_3}; (d) X_T (net reactance) of the $L_2 C_3$ branch.
Ans.   (a) $f_{\text{osc}} = 2.5$ MHz;   (b) $B = 0.05$ or $\frac{1}{20}$;   (c) $X_{L_1} = 1.571$ kΩ, $X_{L_2} = 78.54\,\Omega$, $X_{C_3} = 1.65$ KΩ;
(d) $X_T = 1.571$ kΩ

2.30 Refer again to Fig. 2-31. If C_3 is increased to 154.4 pF, calculate: (a) f_{osc}; (b) X_{L_1}, X_{L_2}, and X_{C_3}; (c) X_T of the $L_2 C_3$ branch.
Ans.   (a) $f_{\text{osc}} = 1.25$ MHz; (b) $X_{L_1} = 785.4\,\Omega$, $X_{L_2} = 39.27\,\Omega$, $X_{C_3} = 824.64\,\Omega$; (c) $X_T = 785.4\,\Omega$

2.31 Refer to Fig. 2-31. Assume that it is desired to obtain an oscillating frequency of 4 MHz. If $C_3 = 25$ pF and B must equal $\frac{1}{30}$, calculate: (a) L_T; (b) L_1 and L_2; (c) X_{L_1}, X_{L2}, and X_{C_3}; (d) X_T (net reactance) of the $L_2 C_3$ branch.
Ans.   (a) $L_T = 63.3\ \mu$H;   (b) $L_1 = 61.3\ \mu$H, $L_2 = 2.04\ \mu$H;   (c) $X_{L_1} = 1.54$ kΩ, $X_{L_2} = 51.3\,\Omega$, $X_{C_3} = 1.59$ kΩ;
(d) $X_T = 1.54$ kΩ

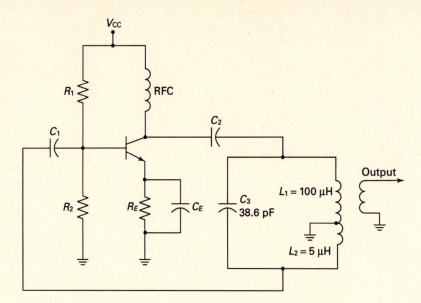

Fig. 2-31

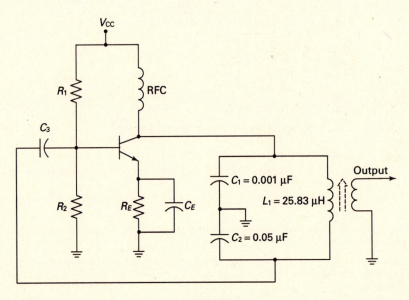

Fig. 2-32

2.32 Refer to Fig. 2-31. If the ac voltage across L_1 is 24 V p–p, how much ac voltage exists at the base of the transistor? (Use original circuit values.)   *Ans.*   1.2 V p–p

2.33 Refer to Fig. 2-32. Calculate the following: (*a*) f_{osc}; (*b*) the feedback fraction B; (*c*) X_{C_1}, X_{C_2}, and X_{L_1}; (*d*) X_T of the L_1C_2 branch.
Ans.   (*a*) $f_{osc} = 1$ MHz; (*b*) $B = 0.02$ or $\frac{1}{50}$; (*c*) $X_{C_1} = 159.2\ \Omega$, $X_{C_2} = 3.18\ \Omega$, $X_{L_1} = 162.3\ \Omega$; (*d*) $X_T = 159.2\ \Omega$

2.34 Refer to Fig. 2-32. Assume that it is desired to obtain an oscillating frequency of 2 MHz and a feedback fraction B of $\frac{1}{20}$. If $L_1 = 50\ \mu$H, calculate the following: (*a*) C_{eq}; (*b*) C_1 and C_2; (*c*) X_{C_1}, X_{C_2} and X_{L_1}; (*d*) X_T (net reactance) of the L_1C_2 branch.
Ans.   (*a*) $C_{eq} = 126.65$ pF;   (*b*) $C_1 = 133$ pF, $C_2 = 2.65$ nF;   (*c*) $X_{C_1} = 598.3\ \Omega$, $X_{C_2} = 30\ \Omega$, $X_{L_1} = 628.3\ \Omega$; (*d*) $X_T = 598.3\ \Omega$

2.35   Refer to Fig. 2-32. If the ac voltage across C_1 equals 21 V p–p, calculate the ac voltage across C_2. (Use original circuit values.)   *Ans.*   420 mV p–p

2.36   What is the phase relationship between the ac voltages across C_1 and C_2?   *Ans.*   180°

2.37   Refer to Fig. 2-33. Calculate the following: (*a*) f_{osc}; (*b*) the feedback fraction *B*; (*c*) X_{C_1}, X_{C_2}, X_{C_3}, and X_{L_1}; (*d*) X_T of the $L_1 C_2 C_3$ branch.
   Ans.   (*a*) $f_{osc} = 5$ MHz; (*b*) $B = 0.025$ or $\frac{1}{40}$; (*c*) $X_{C_1} = 127.3\ \Omega$, $X_{C_2} = 3.18\ \Omega$, $X_{C_3} = 7.726$ kΩ, $X_{L_1} = 7.854$ kΩ;
   (*d*) $X_T = 125\ \Omega$.

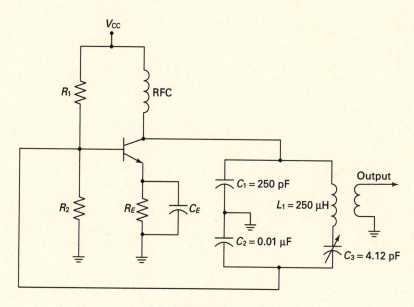

Fig. 2-33

2.38   Refer to Fig. 2-33. Calculate the value of C_3 required to produce an oscillating frequency of 7.5 MHz. Ignore the effects of C_1 and C_2 in determining f_{osc}.   *Ans.* $C_3 = 1.8$ pF

2.39   Refer to Fig. 2-33. Assume $C_1 = 330$ pF, $C_2 = 0.033\ \mu$F, $C_3 = 40.5$ pF, and $L_1 = 50\ \mu$H. Calculate: (*a*) f_{osc}, (*b*) the feedback fraction *B*.
   Ans.   (*a*) $f_{osc} = 3.75$ MHz, (*b*) $B = 0.01$ or $\frac{1}{100}$

2.40   Refer to Fig. 2-34. Calculate the feedback fraction *B*.   *Ans.* $B \approx 0.091$ or $\frac{1}{11}$

2.41   In Fig. 2-34, does the feedback network provide a phase shift for the feedback voltage?
   Ans.   No, the output voltage and feedback voltage are in phase.

2.42   Refer to Fig. 2-35. Calculate the minimum and maximum values for f_N.
   Ans. $f_{N(min)} = 5.94$ kHz, $f_{N(max)} = 303$ kHz

2.43   Refer to Fig. 2-35. To what value must the resistance R_B be adjusted to obtain an f_N of 100 kHz?
   Ans. $R_B = 2.03$ kΩ

2.44   In Fig. 2-35, assume $R_B = 5.06$ kΩ and $C_F = 0.047\ \mu$F. Calculate the following: (*a*) f_N, (*b*) B_L, (*c*) B_C, (*d*) C_{in}.
   Ans.   (*a*) $f_N = 50$ kHz, (*b*) $B_L = 33.33$ kHz, (*c*) $B_C = 7.92$ kHz, (*d*) $C_{in} = 14.5$ nF

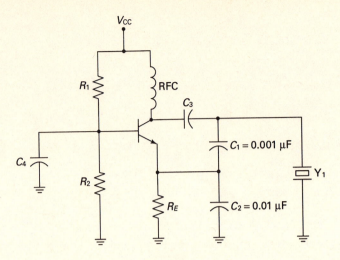

Fig. 2-34

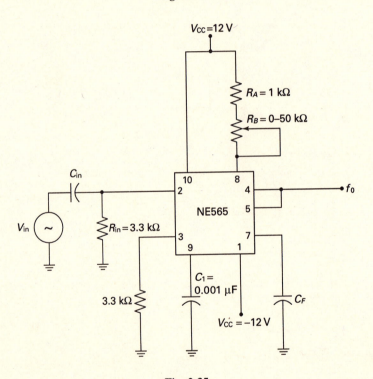

Fig. 2-35

2.45 In Fig. 2-35, assume $R_B = 212 \ \Omega$ and $C_F = 0.0056 \ \mu$F. Calculate the following: (a) f_N, (b) B_L, (c) B_C, (d) C_{in}.
   Ans.   (a) $f_N = 250$ kHz, (b) $B_L = 166.7$ kHz, (c) $B_C = 51.3$ kHz, (d) $C_{in} = 2.89$ nF

2.46 Repeat Problem 2.45 if $V_{CC} = 9$ V.
   Ans.   (a) $f_N = 250$ kHz, (b) $B_L = 222.2$ kHz, (c) $B_C = 59.23$ kHz, (d) $C_{in} = 3.47$ nF

2.47 Refer to Fig. 2-35. Assume $R_B = 3.04$ kΩ, $R_{in} = 8.2$ kΩ, and $C_F = 0.033 \ \mu$F. Calculate the following: (a) f_N, (b) B_L,
   (c) B_C, (d) C_{in}.
   Ans.   (a) $f_N = 75$ kHz, (b) $B_L = 50$ kHz, (c) $B_C = 11.57$ kHz, (d) $C_{in} = 3.88$ nF

2.48 In Problem 2.47, what is the output frequency if the PLL is not in phase lock?
  Ans. $f_o = f_N = 75$ kHz

2.49 In Fig. 2-36, what is f_o if $f_i = 25$ kHz and $N = 100$?   *Ans.* $f_o = 2.5$ MHz

2.50 In Fig. 2-36, what is f_o if $f_i = 100$ kHz and $N = 35$?   *Ans.* $f_o = 3.5$ MHz

2.51 In Fig. 2-37, calculate N_{min} and N_{max}.   *Ans.* $N_{min} = 100$, $N_{max} = 250$

2.52 In Fig. 2-37, what is f_o if: (*a*) $N = 125$, (*b*) $N = 175$, (*c*) $N = 200$, (*d*) $N = 225$?
  Ans. (*a*) $f_o = 43.125$ MHz, (*b*) $f_o = 44.375$ MHz, (*c*) $f_o = 45$ MHz, (*d*) $f_o = 45.625$ MHz

2.53 In Fig. 2-38, what is f_o?   *Ans.* $f_o = 37.5$ MHz

2.54 In Fig. 2-38, assume N is variable between 50 and 75. Calculate: (*a*) $f_{o(min)}$, (*b*) $f_{o(max)}$, (*c*) f_{CH}.
  Ans. (*a*) $f_{o(min)} = 18.75$ MHz, (*b*) $f_{o(max)} = 28.125$ MHz, (*c*) $f_{CH} = 375$ kHz

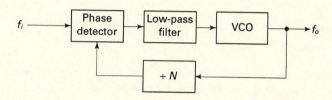

Fig. 2-36

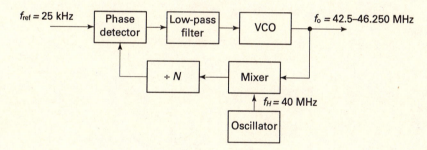

Fig. 2-37

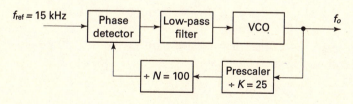

Fig. 2-38

Chapter 3

Amplitude Modulation

INTRODUCTION

Modulation is the process of imposing information contained in a lower-frequency electronic signal onto a higher-frequency signal. The higher-frequency signal is called the *carrier* and the lower-frequency signal is called the *modulating signal*. If the information is imposed on the carrier by causing its amplitude to vary in accordance with the modulating signal, the method is called *amplitude modulation.*

The advantage of transmitting the higher-frequency signal is twofold: First, if all radio stations broadcast simultaneously at audio frequencies, they could not be distinguished from one another and only a jumbled mess would be received. Second, it is found that antennas on the order of magnitude of 5 miles to 5000 miles are necessary for *audio* frequency transmissions.

3.1 MATHEMATICAL DESCRIPTION

The mathematical description of the unmodulated carrier wave is

$$A \sin 2\pi f_c t \tag{3.1}$$

where f_c is the carrier frequency and A is the peak value of the unmodulated carrier.

If, for simplicity, a single audio tone is taken as the modulating signal, it can be represented by

$$B \sin 2\pi f_a t \tag{3.2}$$

where f_a is the frequency of the audio tone and B is the peak value of the modulating signal (see Fig. 3-1).

The modulated wave can be represented mathematically as the product

$$(A + B \sin 2\pi f_a t)(\sin 2\pi f_c t) \tag{3.3}$$

where f_a is the frequency of the audio modulating signal and f_c is the frequency of the carrier. Factoring, we get

$$A\left(1 + \frac{B}{A}\sin 2\pi f_a t\right)(\sin 2\pi f_c t)$$

In terms of voltage, we have

$$v = V_c\left(1 + \frac{B}{A}\sin 2\pi f_a t\right)(\sin 2\pi f_c t) \tag{3.4}$$

where V_c is the peak voltage of the unmodulated carrier, represented by A until now.

Making use of the trigonometric identity

$$(\sin X)(\sin Y) = \tfrac{1}{2}\cos(X - Y) - \tfrac{1}{2}\cos(X + Y)$$

the equation describing the amplitude-modulated wave may be written as

$$\boxed{v = V_c \sin 2\pi f_c t + \frac{mV_c}{2}\cos 2\pi(f_c - f_a)t - \frac{mV_c}{2}\cos 2\pi(f_c + f_a)t} \tag{3.5}$$

where m is called the *modulation factor* and is defined as

$$m = \frac{\text{peak value of modulating signal}}{\text{peak value of unmodulated carrier}} \tag{3.6}$$

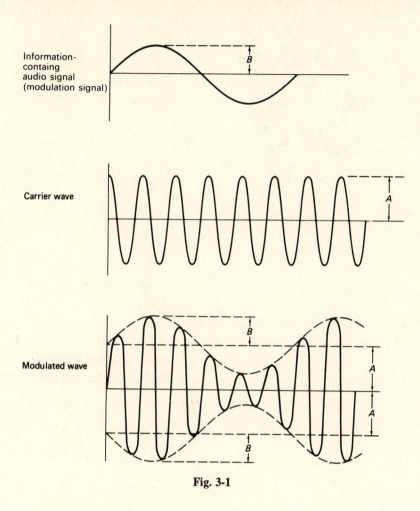

Fig. 3-1

When expressed as a percentage, this is known as the *percent modulation, M*. Using the notation of Fig. 3-1,

$$m = \frac{B}{A}$$

$$M = \frac{B}{A} \times 100\% \tag{3.7}$$

The percent modulation can vary anywhere from 0 to 100% without introducing distortion. If the percent modulation is allowed to increase beyond 100%, distortion, accompanied by undesirable extraneous frequencies, results. Figure 3-2 depicts three degrees of modulation: (*a*) undermodulation ($M < 100\%$), (*b*) 100% modulation, and (*c*) overmodulation ($M > 100\%$).

Referring to the equation above describing the amplitude-modulated wave, the modulated wave is seen to have three components: one at a frequency of f_c, one at a frequency of $f_c + f_a$, and one at a frequency of $f_c - f_a$ producing the frequency-versus-voltage spectrum as shown in Fig. 3-3(*a*).

The frequency $f_c + f_a$ is called the *upper-side frequency*, and $f_c - f_a$ is called the *lower-side frequency*. Most audio information to be broadcast does not consist of a single pure sine wave. In most cases, a rather complex waveshape is encountered. Any complex waveshape can be considered to be the sum of a set of pure sine waves.

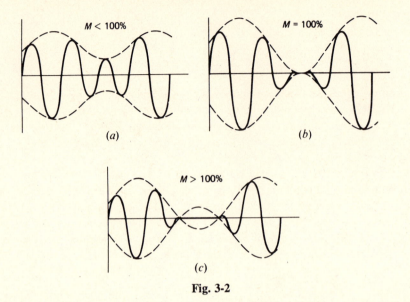

Fig. 3-2

Thus, each of the sine waves which make up the complex audio wave will have both an upper- and a lower-side frequency, which will appear in the analysis of the modulated wave. Rather than discuss an upper- and a lower-side frequency, an upper and a lower sideband of frequencies is referred to.

It can therefore be seen that a broadcast station which intends to broadcast information containing frequencies from 0 to 5000 Hz (5 kHz) needs an upper sideband of 5 kHz and a lower sideband of 5 kHz, for a total bandwidth requirement of 10 kHz. Federal Communications Commission regulations allow a bandwidth of 10 kHz to a station in the AM broadcast band.

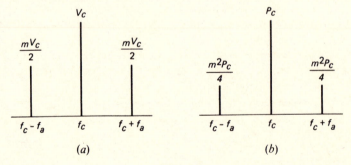

Fig. 3-3

3.2 POWER CONTENT

Since power is proportional to the square of voltage, the power-versus-frequency spectrum for an amplitude-modulated wave looks like Fig. 3-3(b). Each sideband has a power content equal to $m^2 P_c/4$, where P_c is the power content of the signal at the carrier frequency.

Thus the total power is

$$P_T = \frac{m^2 P_c}{4} + \frac{m^2 P_c}{4} + P_c \qquad (3.8)$$

where P_c is the power content of the carrier and is independent of percent modulation in an AM transmission.

Combining terms,

$$P_T = \frac{m^2 P_c}{2} + P_c \qquad (3.9)$$

Factoring,

$$P_T = P_c\left(1 + \frac{m^2}{2}\right) \qquad (3.10)$$

When doing a numerical analysis of power content distribution, it will be found that under optimum conditions (100% modulation) only one-third of the power transmitted is located in the sidebands. Two-thirds of the power is located at the carrier frequency. No information is contained at the carrier frequency. All information is contained within the upper and lower sidebands. In actuality, the two sidebands contain identical information.

Schemes have been devised for making better use of the available power being transmitted. These schemes include suppressed-carrier transmission, double-sideband transmission, and single-sideband transmission. The frequency spectra for these three improved modulation schemes are shown in Fig. 3-4. Essentially, in each of these schemes the power is put where the information is. An additional advantage of the single-sideband scheme is that only half as much bandwidth is required for the transmission, and therefore twice as many stations can transmit simultaneously.

Fig. 3-4

3.3 USING THE OSCILLOSCOPE TO DETERMINE PERCENT MODULATION

Two techniques exist for the determination of percent modulation of an AM wave using the oscilloscope.

In one technique, a standard time base (sawtooth wave) is applied to the horizontal input of the 'scope and the AM wave being examined is put onto the vertical input of the 'scope. The AM wave is then displayed on the 'scope face as shown in Fig. 3-5.

To determine the percent modulation from this oscilloscope display, the difference between the maximum peak-to-peak amplitude and the minimum peak-to-peak amplitude is divided by their sum. Examining this mathematically,

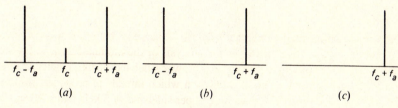

$$\boxed{M = \frac{\max p\text{-}p - \min p\text{-}p}{\max p\text{-}p + \min p\text{-}p} \times 100}$$

$$M = \frac{2(A + B) - 2(A - B)}{2(A + B) + 2(A - B)} \times 100$$

$$= \frac{4B}{4A} \times 100$$

$$= \frac{B}{A} \times 100$$

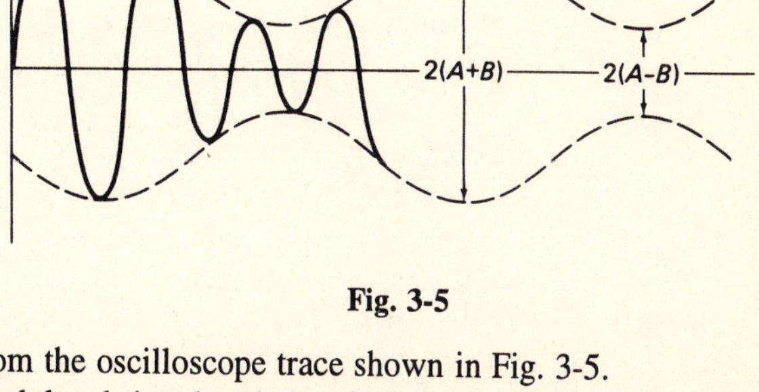

Fig. 3-5

Thus percent modulation can be obtained from the oscilloscope trace shown in Fig. 3-5.

The other technique involves applying the modulated signal to the vertical input of the 'scope and the audio frequency modulating signal to the horizontal input. This results in trapezoidal patterns such as those shown in Fig. 3-6.

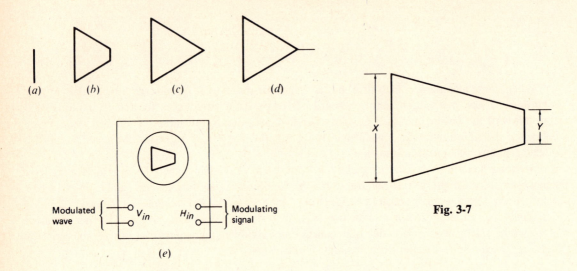

Fig. 3-7

Fig. 3-6

The percent modulation can be obtained from the trapezoidal pattern shown in Fig. 3-7 by using the formula

$$M = \frac{X - Y}{X + Y} \times 100 \qquad (3.11)$$

Sometimes the sides of the trapezoid do not appear to be absolutely straight. If the sides of the trapezoid are nonlinear, this indicates that distortion is present in the output signal. Two examples of nonlinear trapezoid sides are shown in Fig. 3-8.

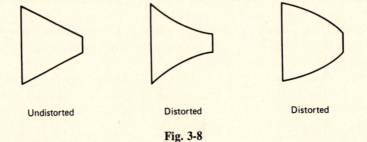

Fig. 3-8

3.4 THE AM TRANSMITTER

Figure 3-9 is a block diagram of an AM transmitter. No matter how complicated an AM transmitter may become, it is basically the same as that shown in Fig. 3-9.

It is necessary to have a nonlinear device in the system in order for modulation to occur, that is, to create the sum and difference frequencies necessary for sidebands to appear.

The nonlinear device in which modulation occurs is the *modulated amplifier*.

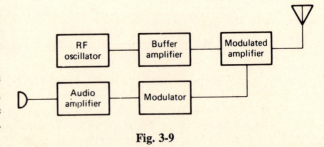

Fig. 3-9

Figure 3-10 shows two rather simple examples of transistor class C amplifiers which can be used as modulated amplifiers. The fact that these amplifiers are class C means that output current is cut off for a

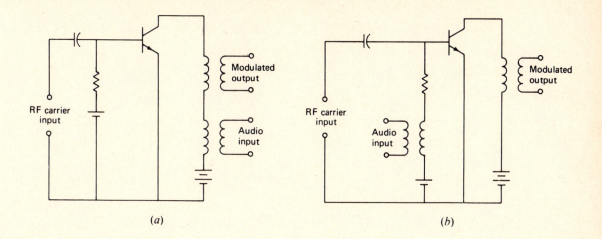

(a)   (b)

Fig. 3-10

portion of the cycle, thereby causing clipping of the output signal. This is then a nonlinear operation and produces the necessary sum and difference frequencies required of an AM transmitter.

Note in particular the terminology used in Fig. 3-9. The modulator is the last stage of audio amplification, while the modulated amplifier is the circuit within which modulation occurs.

3.5 SINGLE-SIDEBAND, DOUBLE-SIDEBAND, AND PILOT-CARRIER TRANSMISSION

In single-sideband transmission, SSB, only one sideband is transmitted, leaving off the other sideband and the carrier. This appreciably reduces the amount of power and bandwidth necessary to be transmitted for a given amount of information.

Another added advantage to SSB over standard AM is that since the signal has a narrower bandwidth a narrower passband is permissible within the receiver, thereby limiting noise pickup because of the narrower open bandwidth.

Double-sideband transmission, DSB, is another variation on AM. In this case, only the carrier is eliminated.

Another variation is called pilot-carrier transmission, in which the two sidebands as well as a trace of the carrier are transmitted.

3.6 THE BALANCED MODULATOR

One means of suppressing a carrier signal in order to create an SSB or DSB signal is to use a circuit known as a *balanced modulator*.

The term "balanced modulator" is an exception to the rule stated earlier in this chapter. At that point it was claimed that a modulator is the last audio amplifier stage. In the case of the balanced modulator, we are actually dealing with the circuitry in which modulation is taking place. It should have been called the "balanced modulated device" if the original rule were followed.

Figure 3-11 shows some variations of balanced modulators.

Although there are many advantages to SSB, DSB, and pilot-carrier transmission, they have not gained general acceptance for use in home-entertainment equipment but are reserved for point-to-point communications because of the complexity of the equipment and thus the associated increased cost. The reason for the extra cost is the need to reconstitute an AM signal prior to demodulation.

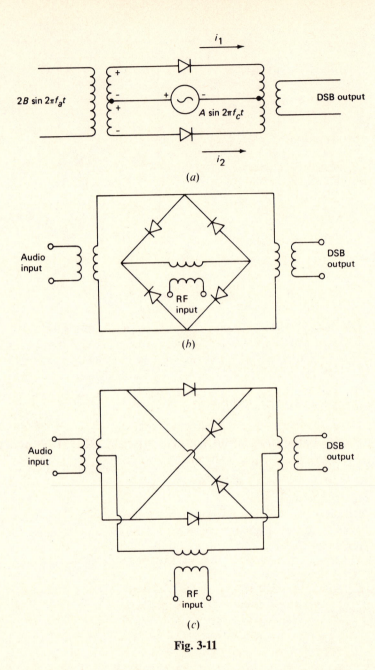

Fig. 3-11

3.7 AM RECEIVERS

The simplest AM receiver consists of a tandem arrangement of a selector–RF amplifier combination, a demodulator, an audio amplifier, and a speaker as shown in Fig. 3-12. This simple receiver is called a tuned radio-frequency (TRF) receiver.

The TRF receiver has been replaced for the most part by the superheterodyne receiver except in cases of the least expensive toylike equipment.

The main difficulty with the TRF receiver is the varying passband encountered as the receiver is tuned from the low end of the frequency band to be received to the high end of the frequency band to be received.

In the superheterodyne receiver, a block diagram of which is shown as Fig. 3-13, most of the high-frequency amplification takes place in the intermediate-frequency section, the passband and center

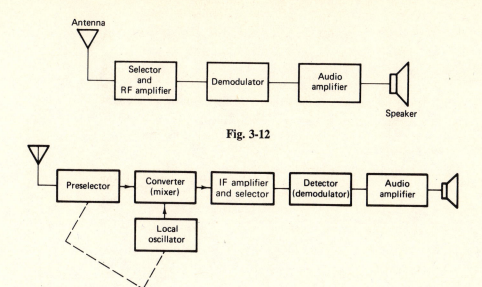

Fig. 3-12

Fig. 3-13

frequency of which stay the same as the receiver is tuned all the way from the lower portion of the band to be received up to the upper end of the band to be received.

The manner in which this can be caused to occur is to heterodyne (beat) the incoming signal with a signal generated by the local oscillator. The local oscillator is tuned simultaneously with the RF selector in such a way that the difference in frequency between the carrier of the station tuned to and the frequency of the local oscillator remains the same. This difference frequency is the intermediate frequency, IF, of the receiver.

The most common intermediate frequency used with commercial AM broadcast receivers is 455 kHz.

Solved Problems

3.1   An audio signal

$$15 \sin 2\pi(1500t)$$

amplitude modulates a carrier

$$60 \sin 2\pi(100,000t)$$

(a) Sketch the audio signal.

(b) Sketch the carrier.

(c) Construct the modulated wave.

(d) Determine the modulation factor and percent modulation.

(e) What are the frequencies of the audio signal and the carrier?

(f) What frequencies would show up in a spectrum analysis of the modulated wave?

SOLUTION

Given: Audio signal $= 15 \sin 2\pi(1500t)$
            Carrier $= 60 \sin 2\pi(100,000t)$

Find: (a) Sketch of audio signal

            (b) Sketch of carrier

            (c) Sketch of modulated wave

            (d) m, M

            (e) f_a, f_c

            (f) Frequency content of modulated wave

(a)   See Fig. 3-14(a).

(b)   See Fig. 3-14(b).

(c)   First develop the envelope of the modulated wave:

   1.   Locate the amplitude of the carrier (dashed line).
   2.   Using the amplitude of the carrier as an axis, lay in the audio signal.

   Now that the envelope has been determined, a signal having an amplitude defined by the envelope found above and having a frequency of the carrier is laid in within the envelope.

   See Fig. 3-14(c).

(d)   Using the following equation for modulation factor,

$$m = \frac{\text{audio amplitude}}{\text{carrier amplitude}} = \frac{B}{A}$$

$$= \tfrac{15}{60} = \tfrac{1}{4}$$

$$\boxed{m = 0.25}$$

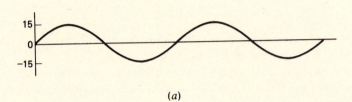

(a)

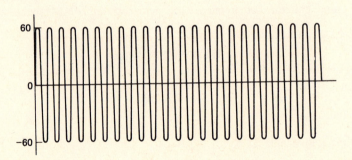

(b)

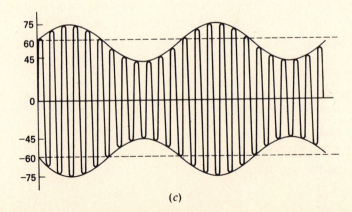

(c)

Fig. 3-14

Converting modulation factor to percent modulation,

$$M = m \times 100$$
$$= 0.25 \times 100$$

$$\boxed{M = 25\%}$$

(e)   Since   Audio signal $= B \sin 2\pi f_a t$
$$= 15 \sin 2\pi(1500t)$$

$$\boxed{f_a = 1500 \text{ Hz}}$$

Since   Carrier $= A \sin 2\pi f_c t$
$$= 60 \sin 2\pi(100,000t)$$

$$\boxed{f_c = 100,000 \text{ Hz}}$$

(f)   The frequency spectrum of an amplitude-modulated wave consists of

$$f_c, \qquad f_c + f_a, \qquad \text{and} \qquad f_c - f_a$$
$$f_c = 100,000 \text{ Hz}$$
$$f_c + f_a = 100,000 + 1500 = 101,500 \text{ Hz}$$
$$f_c - f_a = 100,000 - 1500 = 98,500 \text{ Hz}$$

The frequency content of the modulated wave is

$$\boxed{\begin{array}{l} 100,000 \text{ Hz} \\ 101,500 \text{ Hz} \\ 98,500 \text{ Hz} \end{array}}$$

3.2   A 75-MHz carrier having an amplitude of 50 V is modulated by a 3-kHz audio signal having an amplitude of 20 V.

(a)   Sketch the audio signal.

(b)   Sketch the carrier.

(c)   Construct the modulated wave.

(d)   Determine the modulation factor and percent modulation.

(e)   What frequencies would show up in a spectrum analysis of the modulated wave?

(f)   Write trigonometric equations for the carrier and the modulating waves.

SOLUTION

Given: $f_c = 75 \text{ MHz}$
$A = 50 \text{ V}$
$f_a = 3 \text{ kHz}$
$B = 20 \text{ V}$

Find: (a)   Sketch of audio signal

(b)   Sketch of carrier

(c)   Sketch of modulated wave

(d) m, M

(e)   Frequency content of modulated wave

(f)   Trigonometric equations for modulating signal and carrier

(a)   See Fig. 3-15(a).

(b)   See Fig. 3-15(b).

(c)   The envelope of the carrier is first developed by drawing the horizontal dashed line at the unmodulated carrier amplitude, both positive and negative. The audio signal is now sketched around the dashed line, providing the envelope within which the radio-frequency signal can be laid in. See Fig. 3-15(c).

(d)   From the defining equation of modulation factor,

$$m = \frac{B}{A}$$

$$= \frac{20}{50}$$

$$\boxed{m = 0.4}$$

Percent modulation can now be determined by multiplying the modulation factor by 100:

$$M = m \times 100$$

$$= 0.4 \times 100$$

$$\boxed{M = 40\%}$$

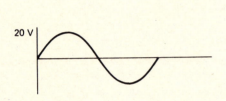

(a)

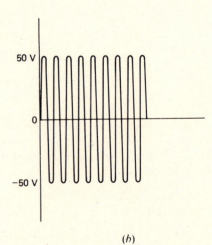

(b)

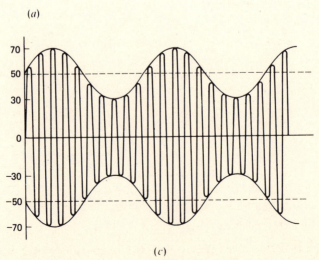

(c)

Fig. 3-15

(e) The frequency content of an AM signal consists of the carrier frequency and the side frequencies which result from adding the audio frequency to the carrier and from subtracting the audio frequency from the carrier frequency.

$$f_c = 75\,\text{MHz}$$
$$f_c + f_a = 75\,\text{MHz} + 3\,\text{kHz}$$
$$= 75{,}000\,\text{kHz} + 3\,\text{kHz}$$
$$= 75{,}003\,\text{kHz}$$
$$f_c - f_a = 75{,}000\,\text{kHz} - 3\,\text{kHz}$$
$$= 74{,}997\,\text{kHz}$$

Thus, the frequency content of the AM wave is

$$\boxed{\begin{array}{l} 75.000\,\text{MHz} \\ 75.003\,\text{MHz} \\ 74.997\,\text{MHz} \end{array}}$$

(f)
$$v_a = B \sin 2\pi f_a t$$
$$= 20 \sin 2\pi (3000) t$$
$$\boxed{v_a = 20 \sin 6000 \pi t}$$
$$v_c = A \sin 2\pi f_c t$$
$$= 50 \sin 2\pi (75 \times 10^6) t$$
$$\boxed{v_c = 50 \sin 150 \times 10^6\, \pi t}$$

(Remember that A represents the amplitude of the carrier and B the amplitude of the modulating signal.)

3.3 How many AM broadcast stations can be accommodated in a 100-kHz bandwidth if the highest frequency modulating a carrier is 5 kHz?

SOLUTION

Given: Total BW = 100 kHz

$f_{a\,\text{max}} = 5\,\text{kHz}$

Find: Number of stations

Any station being modulated by a 5-kHz signal will produce an upper-side frequency 5 kHz above its carrier and a lower-side frequency 5 kHz below its carrier, thereby requiring a bandwidth of 10 kHz. Thus,

$$\text{Number of stations accommodated} = \frac{\text{total BW}}{\text{BW per station}}$$
$$= \frac{100 \times 10^3}{10 \times 10^3}$$

$$\boxed{\text{Number of stations accommodated} = 10 \text{ stations}}$$

3.4 A bandwidth of 20 MHz is to be considered for the transmission of AM signals. If the highest audio frequencies used to modulate the carriers are not to exceed 3 kHz, how many stations could broadcast within this band simultaneously without interfering with one another?

SOLUTION

Given: Total BW $= 20$ MHz

            $f_{a\max} = 3$ kHz

Find: Number of AM stations

The maximum bandwidth of each AM station is determined by the maximum frequency of the modulating signal.

$$\text{Station BW} = 2f_{a\max}$$
$$= 2 \times 3 \times 10^3 = 6 \times 10^3$$
$$= 6 \text{ kHz}$$

Thus, the number of stations that can broadcast simultaneously without interfering with one another is

$$\frac{20 \times 10^6}{6 \times 10^3} = 3.333 \times 10^3$$

> Number of stations $= 3333$

3.5   The total power content of an AM signal is 1000 W. Determine the power being transmitted at the carrier frequency and at each of the sidebands when the percent modulation is 100%.

SOLUTION

Given: $P_T = 1000$ W

            $M = 100\%$; therefore, $m = 1$

Find: P_c, P_{USB}, P_{LSB}

The total power consists of the power at the carrier frequency, that at the upper sideband, and that at the lower sideband.

$$P_T = P_c + P_{\text{USB}} + P_{\text{LSB}}$$

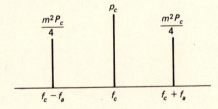

Fig. 3-16

See Fig. 3-16.

From the equation for total power,

$$P_T = P_c + \frac{m^2 P_c}{4} + \frac{m^2 P_c}{4}$$

$$= P_c + \frac{m^2 P_c}{2}$$

$$1000 = P_c + \frac{(1.0)^2 P_c}{2}$$

$$= P_c + 0.5 P_c$$

$$= 1.5 P_c$$

$$\frac{1000}{1.5} = P_c$$

$$P_c = 666.67 \text{ W}$$

This leaves $1000 - 666.67 = 333.33$ W to be shared equally between upper and lower sidebands.

$$P_{\text{USB}} + P_{\text{LSB}} = 333.33 \text{ W}$$

$$P_{\text{USB}} = P_{\text{LSB}}$$

$$2P_{\text{LSB}} = 333.33$$

$$P_{\text{LSB}} = P_{\text{USB}} = \frac{333.33}{2}$$

$$= 166.66$$

$$P_c = 666.67 \text{ W}$$
$$P_{\text{USB}} = P_{\text{LSB}} = 166.66 \text{ W}$$

3.6 Determine the power content of the carrier and each of the sidebands for an AM signal having a percent modulation of 80% and a total power of 2500 W.

SOLUTION

Given: $M = 80\%$; $m = 0.8$
          $P_T = 2500 \text{ W}$

Find: P_c, P_{USB}, P_{LSB}

The total power of an AM signal is the sum of the power at the carrier frequency and the power contained in the sidebands.

$$P_T = P_c + P_{\text{USB}} + P_{\text{LSB}}$$

Using the equation for total power,

$$P_T = P_c + \frac{m^2 P_c}{4} + \frac{m^2 P_c}{4}$$

$$\frac{m^2 P_c}{4} + \frac{m^2 P_c}{4} = \frac{m^2 P_c}{2}$$

$$P_T = P_c + \frac{m^2 P_c}{2}$$

$$2500 = P_c + \frac{(0.8)^2 P_c}{2}$$

$$= P_c + \frac{0.64}{2} P_c = 1.32 P_c$$

$$= 1.32 P_c$$

$$P_c = \frac{2500}{1.32}$$

$$\boxed{P_c = 1893.9 \text{ W}}$$

The power in the two sidebands is the difference between the total power and the carrier power.

$$P_{\text{USB}} + P_{\text{LSB}} = 2500 - 1893.9$$
$$P_{\text{USB}} + P_{\text{LSB}} = 606.1 \text{ W}$$

$$P_{\text{USB}} = P_{\text{LSB}} = \frac{606.1}{2} \text{ W}$$

$$\boxed{P_{\text{USB}} = P_{\text{LSB}} = 303.50 \text{ W}}$$

3.7 The power content of the carrier of an AM wave is 5 kilowatts (kW). Determine the power content of each of the sidebands and the total power transmitted when the carrier is modulated 75%.

SOLUTION

Given: $P_c = 5 \text{ kW}$
          $M = 75\%$; $m = 0.75$

Find: P_{USB}, P_{LSB}, P_T

Since, in an AM wave, the power in each of the sidebands is equal,

$$P_{\text{USB}} = P_{\text{LSB}} = \frac{m^2 P_c}{4}$$

$$= \frac{(0.75)^2 (5000)}{4}$$

$$\boxed{P_{\text{USB}} = P_{\text{LSB}} = 703.13 \text{ W}}$$

The total power is the sum of the carrier power and the power in the two sidebands.

$$P_T = P_c + P_{\text{USB}} + P_{\text{LSB}}$$

$$= 5000 + 703.13 + 703.13$$

$$\boxed{P_T = 6406.26 \text{ W}}$$

3.8   An amplitude-modulated wave has a power content of 800 W at its carrier frequency. Determine the power content of each of the sidebands for a 90% modulation.

SOLUTION

Given: $P_c = 800$ W

$M = 90\%$; $m = 0.90$

Find: P_{LSB}, P_{USB}

The power in each of the sidebands is equal to $m^2 P_c/4$.

$$P_{\text{LSB}} = P_{\text{USB}} = \frac{m^2 P_c}{4}$$

$$= \frac{(0.9)^2 \, 800}{4}$$

$$\boxed{P_{\text{LSB}} = P_{\text{USB}} = 162 \text{ W}}$$

3.9   Determine the percent modulation of an amplitude-modulated wave which has a power content at the carrier of 8 kW and 2 kW in each of its sidebands when the carrier is modulated by a simple audio tone.

SOLUTION

Given: $P_c = 8$ kW

$P_{\text{USB}} = P_{\text{LSB}} = 2$ kW

Find: M

Knowing the power content of the sidebands and the carrier, the relationship of sideband power can be used to determine the modulation factor. Once the modulation factor is known, merely multiplying it by 100 provides percent modulation.

$$P_{\text{USB}} = P_{\text{LSB}} = \frac{m^2 P_c}{4}$$

$$2 \times 10^3 = \frac{m^2 (8 \times 10^3)}{4}$$

$$m^2 = \frac{4 \times 2 \times 10^3}{8 \times 10^3}$$

$$= 1.0$$

$$m = 1.0$$

$$M = m \times 100$$

$$\boxed{M = 100\%}$$

3.10 The total power content of an AM wave is 600 W. Determine the percent modulation of the signal if each of the sidebands contains 75 W.

SOLUTION

Given: $P_T = 600$ W

$\quad\quad\quad\quad P_{\text{USB}} = P_{\text{LSB}} = 75$ W

Find: M

In order to determine the percent modulation, the power contained at the carrier frequency is first determined. Once P_c is known, the relationship between P_c and the sideband power will provide a means of determining the modulation factor, from which the percent modulation is easily found.

Carrier power can be determined from the following:

$$P_T = P_c + P_{\text{USB}} + P_{\text{LSB}}$$

$$600 = P_c + 75 + 75$$

$$P_c = 600 - 150$$

$$= 450$$

Now using the relationship between sideband power and carrier power,

$$P_{\text{USB}} = P_{\text{LSB}} = \frac{m^2 P_c}{4}$$

$$75 = \frac{m^2(450)}{4}$$

$$m^2 = \frac{4(75)}{450}$$

$$= 0.667$$

$$m = 0.816$$

Converting modulation factor to percent modulation,

$$M = m \times 100$$

$$= 0.816 \times 100$$

$$\boxed{M = 81.6\%}$$

3.11 Find the percent modulation of an AM wave whose total power content is 2500 W and whose sidebands each contain 400 W.

SOLUTION

Given: $P_T = 2500$ W

$\quad\quad\quad\quad P_{\text{USB}} = P_{\text{LSB}} = 400$ W

Find: M

First find the power contained at the carrier frequency. Then use the relationship between sideband power and carrier power to determine the modulation factor. Once the modulation factor is known, the percent modulation can easily be found merely by multiplying by 100.

The power at the carrier frequency can be determined from the following:

$$P_T = P_c + P_{USB} + P_{LSB}$$

$$2500 = P_c + 400 + 400$$

$$P_c = 2500 - 800$$

$$= 1700 \text{ W}$$

$$P_{USB} = P_{LSB} = \frac{m^2 P_c}{4}$$

$$400 = \frac{m^2(1700)}{4}$$

$$m^2 = \frac{400(4)}{1700}$$

$$= \frac{1600}{1700}$$

$$= 0.941$$

$$m = 0.970$$

$$M = 0.970 \times 100$$

$$\boxed{M = 97\%}$$

3.12   Determine the power content of each of the sidebands and of the carrier of an AM signal that has a percent modulation of 85% and contains 1200 W total power.

SOLUTION

Given: $M = 85\%$; $m = 0.85$

         $P_T = 1200 \text{ W}$

Find: P_c, P_{USB}, P_{LSB}

Using the equation which relates total power to carrier power,

$$P_T = P_c\left(1 + \frac{m^2}{2}\right)$$

$$1200 = P_c\left[1 + \frac{(0.85)^2}{2}\right]$$

$$= P_c\left[1 + \frac{0.7225}{2}\right]$$

$$= P_c[1 + 0.3613]$$

$$= 1.3613 P_c$$

$$P_c = \frac{1200}{1.3613}$$

$$\boxed{P_c = 881.5 \text{ W}}$$

The sum of carrier power and sideband power is equal to total power.

$$P_c + P_{SB} = P_T$$
$$881.5 + P_{SB} = 1200$$
$$P_{SB} = 1200 - 881.5$$
$$= 318.5$$

The total sideband power is made up equally of upper sideband power and lower sideband power.

$$P_{USB} = P_{LSB} = \frac{P_{SB}}{2}$$
$$= \frac{318.5}{2}$$
$$= 159.25$$

$$\boxed{\begin{array}{l} P_c = 881.5 \text{ W} \\ P_{USB} = P_{LSB} = 159.25 \text{ W} \end{array}}$$

3.13 An AM signal in which the carrier is modulated 70% contains 1500 W at the carrier frequency. Determine the power content of the upper and lower sidebands for this percent modulation. Calculate the power at the carrier and the power content of each of the sidebands when the percent modulation drops to 50%.

SOLUTION

Given: $M = 70\%$; $m = 0.70$

$P_{c_{70}} = 1500$ W

Find: $P_{USB_{70}}$, $P_{LSB_{70}}$, $P_{c_{50}}$, $P_{USB_{50}}$, $P_{LSB_{50}}$

The power content of each of the sidebands is equal to $m^2 P_c/4$.

$$P_{USB_{70}} = P_{LSB_{70}} = \frac{m^2 P_c}{4}$$
$$= \frac{(0.7)^2 1500}{4}$$
$$= 183.75$$

$$\boxed{P_{USB_{70}} = P_{LSB_{70}} = 183.75 \text{ W}}$$

In standard AM transmission, carrier power remains the same, regardless of percent modulation. Thus,

$$P_{c_{50}} = P_{c_{70}} = 1500 \text{ W}$$
$$P_{USB_{50}} = P_{LSB_{50}} = \frac{m^2 P_c}{4}$$
$$= \frac{(0.5)^2 1500}{4}$$
$$= 93.75 \text{ W}$$

$$\boxed{\begin{array}{l} P_{c_{50}} = 1500 \text{ W} \\ P_{USB_{50}} = P_{LSB_{50}} = 93.75 \text{ W} \end{array}}$$

3.14 The percent modulation of an AM wave changes from 40% to 60%. Originally, the power content at the carrier frequency was 900 W. Determine the power content at the carrier frequency and within each of the sidebands after the percent modulation has risen to 60%.

SOLUTION

Given: $\quad M_1 = 40\%; \quad m_1 = 0.40$

$\qquad\qquad M_2 = 60\%; \quad m_2 = 0.60$

$\qquad\qquad P_{c_{40}} = 900 \text{ W}$

Find: $\qquad P_{c_{60}}, \quad P_{\text{USB}_{60}}, \quad P_{\text{LSB}_{60}}$

The power content of the carrier of an AM signal remains the same regardless of percent modulation. Thus,

$$P_{c_{60}} = P_{c_{40}} = 900 \text{ W}$$

The power content of each of the sidebands is equal to $m^2 P_c/4$.

$$P_{\text{USB}_{60}} = P_{\text{LSB}_{60}} = \frac{m^2 P_c}{4}$$

$$= \frac{(0.60)^2 (900)}{4}$$

$$\boxed{P_{\text{USB}_{60}} = P_{\text{LSB}_{60}} = 81.0 \text{ W}}$$

3.15 A single-sideband (SSB) signal contains 1 kW. How much power is contained in the sidebands and how much at the carrier frequency?

SOLUTION

Given: $\quad P_{\text{SSB}} = 1 \text{ kW}$

Find: $\qquad P_{\text{SB}}, \quad P_c$

In a single-sideband transmission, the carrier and one of the sidebands have been eliminated. Therefore, all the transmitted power is transmitted at one of the sidebands regardless of percent modulation. Thus,

$$\boxed{\begin{array}{l} P_{\text{SB}} = 1 \text{ kW} \\ P_c = 0 \text{ W} \end{array}}$$

3.16 An SSB transmission contains 10 kW. This transmission is to be replaced by a standard amplitude-modulated signal with the same power content. Determine the power content of the carrier and each of the sidebands when the percent modulation is 80%.

SOLUTION

Given: $\quad P_{\text{SSB}} = 10 \text{ kW}$

$\qquad\qquad M = 80\%; \quad m = 0.80$

Find: $\qquad P_c, \quad P_{\text{LSB}}, \quad P_{\text{USB}}$

Since the total power content of the new AM signal is to be the same as the total power content of the SSB signal,

$$P_T = P_{\text{SSB}} = 10 \text{ kW}$$

Solving for power contained at the carrier frequency,

$$P_T = P_c + P_{\text{LSB}} + P_{\text{USB}}$$

$$= P_c + \frac{m^2 P_c}{4} + \frac{m^2 P_c}{4}$$

$$10{,}000 = P_c + \frac{(0.8)^2 P_c}{4} + \frac{(0.8)^2 P_c}{4}$$

$$= P_c + \frac{0.64 P_c}{2}$$

$$= 1.32 P_c$$

$$\frac{10{,}000}{1.32} = P_c$$

$$P_c = 7575.76 \, \text{W}$$

The power content of the sidebands is equal to the difference between the total power and the carrier power.

$$P_{\text{SB}} = P_T - P_c$$

The power content of the upper and the lower sidebands is equal.

$$P_{\text{LSB}} + P_{\text{USB}} = 10{,}000 - 7575.76$$

$$= 2424.24$$

$$P_{\text{LSB}} = P_{\text{USB}} = \frac{2424.24}{2}$$

$$= 1212.12 \, \text{W}$$

Thus,

$$\boxed{\begin{aligned} P_c &= 7575.76 \, \text{W} \\ P_{\text{LSB}} &= P_{\text{USB}} = 1212.12 \, \text{W} \end{aligned}}$$

3.17 Determine the modulation factor and percent modulation of the signal shown as Fig. 3-17.

SOLUTION

Given: AM signal as shown in Fig. 3-17

Find: m and M

Using the equation relating maximum peak-to-peak amplitude and minimum peak-to-peak amplitude to modulation factor

$$m = \frac{\max p\text{-}p - \min p\text{-}p}{\max p\text{-}p + \min p\text{-}p}$$

we get from Fig. 3-17

$$\max p\text{-}p = 2(80) = 160$$

$$\min p\text{-}p = 2(20) = 40$$

$$m = \frac{160 - 40}{160 + 40} = \frac{120}{200} = 0.6$$

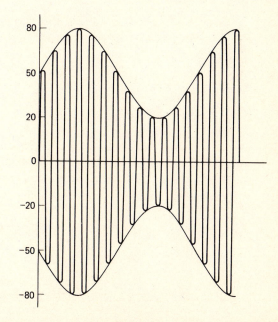

Fig. 3-17

$$\boxed{m = 0.6}$$

$$M = m \times 100 = 0.6 \times 100$$

$$\boxed{M = 60\%}$$

3.18 Find the modulation index and percent modulation of the signal shown as Fig. 3-18.

SOLUTION

Given: AM signal as shown in Fig. 3-18

Find: m and M

Using $m = \dfrac{\max p\text{-}p - \min p\text{-}p}{\max p\text{-}p + \min p\text{-}p}$

and values from Fig. 3-18, we get

$$\max p\text{-}p = 2(50) = 100$$
$$\min p\text{-}p = 2(15) = 30$$
$$m = \frac{100 - 30}{100 + 30} = \frac{70}{130}$$

$$\boxed{m = 0.538}$$

$$M = m \times 100$$

$$\boxed{M = 53.8\%}$$

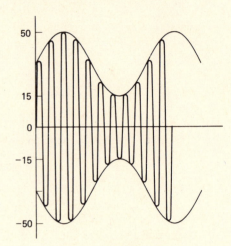

Fig. 3-18

3.19 The trapezoidal pattern shown in Fig. 3-19 results when examining an AM wave. Determine the modulation factor and percent modulation of the wave. What can be said about the distortion of the AM wave?

SOLUTION

Given: Trapezoidal pattern shown as Fig. 3-19

Find: m, M, distortion

Using the equation for percent modulation,

$$M = \frac{x - y}{x + y} \times 100$$

and substituting values from Fig. 3-19, we get

$$M = \frac{5 - 2}{5 + 2} \times 100$$

$$= \frac{3}{7} \times 100$$

$$\boxed{\begin{aligned} M &= 42.9\% \\ m &= 0.429 \end{aligned}}$$

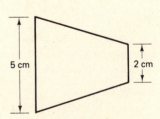

Fig. 3-19

Regarding distortion: Since the sides of the trapezoidal pattern show very little, if any, curvature, it can be said that there is very little, if any, distortion of the modulated wave.

3.20   Determine the modulation factor and percent modulation of the modulated wave which generates the trapezoidal pattern shown as Fig. 3-20.

SOLUTION

Given: Trapezoidal pattern of Fig. 3-20

Find: m and M

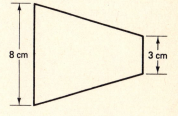

Fig. 3-20

Using

$$M = \frac{x - y}{x + y} \times 100$$

and substituting values from Fig. 3-20, we get

$$M = \frac{8 - 3}{8 + 3} \times 100 = \frac{5}{11} \times 100$$

$$\boxed{M = 45.5\%}$$

The relationship between percent modulation and modulation index is

$$M = m \times 100$$

Substituting numerical values and solving,

$$45.5 = m \times 100$$

$$\frac{45.5}{100} = m$$

$$\boxed{m = 0.455}$$

3.21   An AM standard broadcast receiver is to be designed having an intermediate frequency (IF) of 455 kHz.

(a)   Calculate the required frequency that the local oscillator should be at when the receiver is tuned to 540 kHz if the local oscillator tracks above the frequency of the received signal.

(b)   Repeat (a) if the local oscillator tracks below the frequency of the received signal.

SOLUTION

Given: $f_{IF} = 455$ kHz
            $f_c = 540$ kHz

Find: (a) $f_{LO\,above}$ (b) $f_{LO\,below}$

The intermediate frequency is generated by producing a difference frequency between the carrier and the local oscillator. Thus,

$$f_{IF} = f_c - f_{LO}$$

or

$$f_{IF} = f_{LO} - f_c$$

(a) $$f_{IF} = f_{LO} - f_c$$

    Solving for f_{LO},

$$f_{LO} = f_{IF} + f_c$$
$$= (455 \times 10^3) + (540 \times 10^3)$$

$$\boxed{f_{LO} = 995 \text{ kHz}}$$

(b) $$f_{IF} = f_c - f_{LO}$$

    Solving for f_{LO},

$$f_{LO} = f_c + f_{IF}$$
$$= (540 \times 10^3) + (455 \times 10^3)$$

$$\boxed{f_{LO} = 85 \text{ kHz}}$$

Supplementary Problems

3.22   Why don't broadcast stations transmit at audio frequencies?
    Ans.   Stations could not be distinguished from each other. Antennas would have to be tremendously large.

3.23   Which of the AM waves in Fig. 3-21 depicts undermodulation?   100% modulation?   overmodulation?
    Ans.   under: (a), (e); 100%: (c); over: (b), (d)

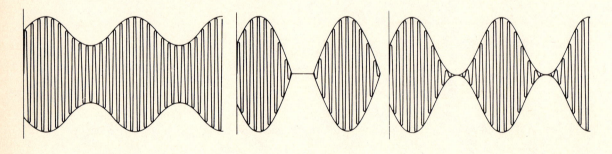

(a)   (b)   (c)

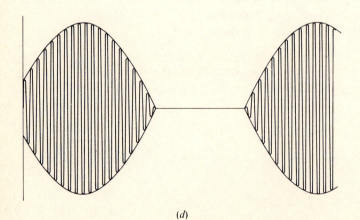

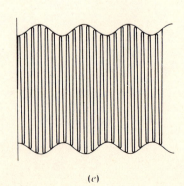

(d)   (e)

Fig. 3-21

3.24   An audio signal whose mathematical description is

$$25 \sin (2\pi 1000t)$$

modulates a carrier described as

$$75 \sin (2\pi 150,000t)$$

(a)   Sketch the audio signal.

(b)   Sketch the carrier.

(c)   Construct the modulated wave showing all amplitude magnitudes.

(d)   Calculate the modulation factor and percent modulation.

(e)   What is the frequency of the audio signal?   of the carrier?

(f)   What frequencies would show up in a spectrum analysis of the modulated wave?

Ans.   (d) 0.333, 33.33%; (e) 1000 Hz, 150,000 Hz; (f) 149,000 Hz, 150,000 Hz, 151,000 Hz

3.25   An audio signal described as

$$30 \sin (2\pi 2500t)$$

amplitude modulates a carrier which is described as

$$65 \sin (2\pi 250,000t)$$

(a)   Sketch the audio signal.

(b)   Sketch the carrier.

(c)   Construct the modulated wave.

(d)   What is the modulation factor and percent modulation?

(e)   What is the frequency of the audio signal?   of the carrier?

(f)   What frequencies would show up in a spectrum analysis of the modulated wave?

Ans.   (d) 0.4615, 46.15%; (e) 2500 Hz, 250,000 Hz; (f) 247,500 Hz, 250,000 Hz, 252,500 Hz

3.26   A 2000-Hz audio signal having an amplitude of 15 V amplitude modulates a 100-kHz carrier which has a peak value of 25 V when not modulated.

(a)   Sketch the audio signal to scale.

(b)   Sketch the carrier to scale.

(c)   Construct the modulated wave to scale.

(d)   Calculate the modulation factor and percent modulation of the modulated wave.

(e)   What frequencies appear in a spectrum analysis of the modulated wave?

Ans.   (d) 0.60, 60%; (e) 98 kHz, 100 kHz, 102 kHz

3.27   An 1800-Hz signal which has an amplitude of 30 V amplitude modulates a 50-MHz carrier which when unmodulated has an amplitude of 65 V.

(a)   Sketch the modulating signal.

(b)   Sketch the carrier.

(c)   Construct the modulated wave.

(d)   Calculate the modulation factor and percent modulation.

(e)   What frequencies would show up in a spectrum analysis of the AM wave?

(f)   Write the trigonometric equations for the carrier and for the audio signal.

Ans.   (d) 0.4615, 46.15%; (e) 49.9982 MHz, 50.0000 MHz, 50.0018 MHz; (f) $65 \sin [2\pi(50 \times 10^6)t]$,
$30 \sin [2\pi(1800)t]$

3.28   How many AM broadcast stations can be accommodated in a 6-MHz bandwidth if each station transmits a signal which was modulated by an audio signal having a maximum frequency of 5 kHz?   *Ans.*   600 stations

3.29   A bandwidth of 12 MHz becomes available for assignment. If assigned for TV broadcast service, only two channels could be accommodated. Determine the number of AM stations that could broadcast simultaneously if the maximum modulating frequency is limited to 5 kHz.   *Ans.*   1200 stations

3.30   A 90-kHz bandwidth is to accommodate six AM broadcasts simultaneously. What maximum modulating frequency must each station be limited to?   *Ans.*   7500 Hz

3.31   An antenna transmits an AM signal having a total power content of 15 kW. Determine the power being transmitted at the carrier frequency and at each of the sidebands when the percent modulation is 85%.
Ans.   11,019W, 1990 W

3.32   Calculate the power content of the carrier and of each of the sidebands of an AM signal whose total broadcast power is 40 kW when the percent modulation is 60%.   *Ans.*   33,898 W, 3050 W

3.33   A 3500-Hz audio tone amplitude modulates a 200-kHz carrier resulting in a modulated signal having a percent modulation of 85%. The total power being transmitted is 15 kW.

   (*a*)   What frequencies would appear in a spectrum analysis of the modulated wave?

   (*b*)   Determine the power content at each of the frequencies that appear in a spectrum analysis of the modulated wave.

   Ans.   (*a*) 196,500 Hz, 200,000 Hz, 203,500 Hz; (*b*) 11,019 W, 1990 W

3.34   Determine the power contained at the carrier frequency and within each of the sidebands for an AM signal whose total power content is 15 kW when the modulation factor is 0.70.   *Ans.*   12,048 W, 1475 W

3.35   An amplitude-modulated signal contains a total of 6 kW. Calculate the power being transmitted at the carrier frequency and at each of the sidebands when the percent modulation is 100%.   *Ans.*   4000 W, 1000 W

3.36   An AM wave has a power content of 1800 W at its carrier frequency. What is the power content of each of the sidebands when the carrier is modulated 85%?   *Ans.*   325 W

3.37   An AM signal contains 500 W at its carrier frequency and 100 W in each of its sidebands.

   (*a*)   Determine the percent modulation of the AM signal.

   (*b*)   Find the allocation of power if the percent modulation is changed to 60%.

   NOTE: The power content of the carrier of an AM wave does not vary with percent modulation.
   Ans.   (*a*) 89.44%; (*b*) 500 W, 45 W, 45 W

3.38   1200 W is contained at the carrier frequency of an AM signal. Determine the power content of each of the sidebands for each of the following percent modulations: (*a*) 40%, (*b*) 50%, (*c*) 75%, (*d*) 100%.
   NOTE: The power content of the carrier of an AM wave does not vary with percent modulation.
   Ans.   (*a*) 48 W, (*b*) 75 W, (*c*) 168.75 W, (*d*) 300 W

3.39   An AM wave has a total transmitted power of 4 kW when modulated 85%. How much total power should an SSB wave contain in order to have the same power content as that contained in the two sidebands?   *Ans.*   1061.52 W

3.40   An SSB transmission contains 800 W. This transmission is to be replaced by a standard AM signal with the same power content. Determine the power content of the carrier and each of the sidebands when the percent modulation is 85%.   *Ans.*   587.695 W, 106.15 W

3.41 Calculate the modulation factor and percent modulation of the AM wave shown as Fig. 3-22.
Ans. 0.50, 50%

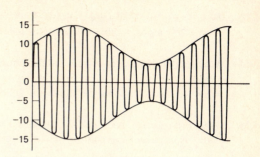

Fig. 3-22

3.42 An AM signal is examined on an oscilloscope. It has a maximum peak to peak of 4.5 cm and a minimum peak to peak of 2 cm.

   (*a*) Sketch the pattern observed on the 'scope.
   (*b*) Determine the modulation factor and percent modulation of the signal.
   (*c*) Calculate the power content of each of the sidebands if the power contained by the signal at the carrier frequency is 500 W.

   Ans. (*b*) 0.3846, 38.46%; (*c*) 18.49 W

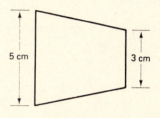

3.43 The trapezoidal pattern shown in Fig. 3-23 results when examining an AM wave. Calculate the modulation factor and percent modulation. *Ans.* 0.25, 25%

Fig. 3-23

3.44 Determine the modulation factor and percent modulation of the AM wave which produces the trapezoidal pattern shown in Fig. 3-24. *Ans.* 0.7143, 71.43%

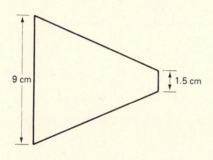

Fig. 3-24

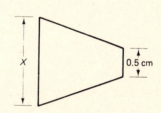

Fig. 3-25

3.45 The trapezoidal pattern of an amplitude-modulated wave having an 85% modulation is shown in Fig. 3-25. Determine the length x of the larger vertical side. *Ans.* 6.167 cm

3.46 Why is it necessary to have a nonlinear device in an AM transmitter?
Ans. To generate sum and difference frequencies.

3.47 Where does modulation occur in most AM transmitters? *Ans.* In the modulated amplifier.

3.48 Why are class C amplifiers suited for use as modulated amplifiers in AM transmitters?
Ans. They are nonlinear.

3.49 What part of an AM transmitter is usually called the modulator?
Ans. The last stage of audio amplification.

3.50 How does single sideband differ from standard AM? *Ans.* It is missing the carrier and one sideband.

3.51 How does double sideband differ from standard AM? *Ans.* It is missing the carrier.

3.52   How does pilot-carrier transmission differ from standard AM transmission?
Ans.   Only a small portion of the carrier is transmitted.

3.53   What are the advantages of SSB and DSB over AM?   *Ans.*   The power is placed where the information is.

3.54   What is an added advantage of SSB over DSB?   *Ans.*   Less noise pickup.

3.55   What is a balanced modulator? Why is this a poor choice of name for this device?
Ans.   A modulator that puts out only two sidebands and no carrier.

3.56   Why have SSB and DSB not found acceptance for use in most home-type entertainment equipment?
Ans.   Cost.

3.57   What function is served by having an intermediate-frequency section in an AM superheterodyne receiver?
Ans.   The passband of the receiver stays the same as the receiver is tuned from the lower end of the spectrum to the upper end.

3.58   What is a local oscillator and what is its function?
Ans.   An oscillator within the receiver which generates a signal to be heterodyned with the incoming signal.

3.59   An AM receiver is tuned to a station whose carrier frequency is 750 kHz. What frequency should the local oscillator be set to in order to provide an intermediate frequency of 455 kHz if the local oscillator frequency tracks below the received frequency?   If it tracks above?   *Ans.*   295 kHz, 1205 kHz

3.60   Repeat Problem 3.59 for a station having a carrier frequency of 1200 kHz.   *Ans.*   745 kHz, 1655 kHz

Chapter 4

Frequency Modulation

INTRODUCTION

Frequency modulation was originally developed to cope with undesirable noise which competed with the desired signal when amplitude modulation was used. Most noise appeared as an additional amplitude modulation on the signal.

When frequency modulating a carrier, information is placed on the carrier by varying its frequency while holding its amplitude fixed. Upon being received, variations in amplitude are eliminated prior to demodulation without affecting the information content contained in the frequency variations, thereby eliminating any noise which may appear as an amplitude modulation of the carrier. See Fig. 4-1.

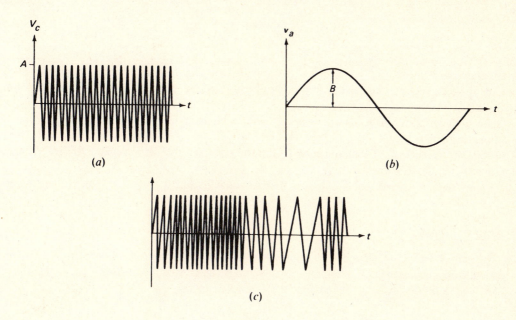

Fig. 4-1

The unmodulated carrier is described as

$$v_c = A \sin 2\pi f t \qquad (4.1)$$

The modulating or audio signal is described as

$$v_a = B \sin 2\pi f_a t \qquad (4.2)$$

The carrier frequency f will vary around a resting frequency f_c thus:

$$f = f_c + \Delta f \sin 2\pi f_a t \qquad (4.3)$$

The frequency-modulated wave will have the following description:

$$v = A \sin [2\pi(f_c + \Delta f \sin 2\pi f_a t)t] \qquad (4.4)$$

In this frequency-modulated situation, Δf is the maximum change in frequency the modulated wave undergoes; it is called the *frequency deviation*. The total variation in frequency, from the lowest to the highest,

111

is referred to as the *carrier swing*. Thus, for a modulating signal which has equal positive and negative peaks, such as a pure sine wave, the carrier swing is equal to two times the frequency deviation.

$$\Delta f = \text{frequency deviation}$$

$$\text{Carrier swing} = 2 \times \text{frequency deviation}$$

It can be shown that the equation for the frequency-modulated wave can be manipulated into

$$v = A \sin\left(2\pi f_c t + \frac{\Delta f}{f_a} \cos 2\pi f_a t\right) \tag{4.5}$$

However, since the mathematics involved in developing this second equation depends on calculus, it will not be shown.

Note that in this equation the cosine term is preceded by the term $\Delta f/f_a$. This quantity is called the *modulation index* and is indicated as m_f.

$$\text{Modulation index} = m_f = \frac{\Delta f}{f_a} \tag{4.6}$$

where Δf is the frequency deviation.

The Federal Communications Commission of the United States requires that frequency modulation be used as the modulation technique for the band of frequencies between 88 MHz and 108 MHz. This is known as the *FM broadcast band*.

Frequency modulation is also mandated as the required modulation technique for the audio portion of the television broadcast band.

The Federal Communications Commission sets a maximum frequency deviation of 75 kHz for FM broadcast stations in the 88- to 108-MHz band.

A maximum frequency deviation of 25 kHz is permitted for the sound portion of television broadcasts.

4.1 PERCENT MODULATION

The term "percent modulation" as it is used in reference to FM refers to the ratio of actual frequency deviation to the maximum allowable frequency deviation. Thus, 100% modulation corresponds to 75 kHz for the commercial FM broadcast band and 25 kHz for television.

$$\text{Percent modulation } M = \frac{\Delta f_{\text{actual}}}{\Delta f_{\text{max}}} \times 100 \tag{4.7}$$

4.2 SIDEBANDS

Analyzing a frequency-modulated wave results in finding that unlike the amplitude-modulated wave, which has only two side frequencies for each modulating frequency, the FM signal has an infinite number of side frequencies spaced f_a apart on both sides of the resting frequency. See Fig. 4-2. Happily, however, most of the side frequencies do not contain significant amounts of power.

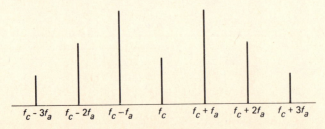

Fig. 4-2

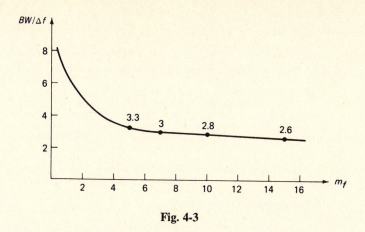

Fig. 4-3

Fourier analysis indicates that the number of side frequencies which contain a significant amount of power, and thus the effective bandwidth of the FM signal, is dependent on the modulation index of the modulated wave, $\Delta f/f_a$.

Schwartz* developed a graph for determining the bandwidth of an FM signal if the modulation index is known. This graph has been reproduced as Fig. 4-3. Schwartz uses as his criterion the rule of thumb that any component frequency with a signal strength (voltage) less than 1% of that of the unmodulated carrier shall be considered too small to be significant.

The Federal Communications Commission sets a 15-kHz maximum limit on the frequency of the modulating signal for both the FM broadcast band and commercial television band (the sound portion of broadcast television transmission is a frequency-modulated signal).

$$f_{a,\,\text{max}} = 15\,\text{kHz} \qquad \text{for both 88 to 108 MHz and TV}$$

4.3 CENTER FREQUENCY AND BANDWIDTH ALLOCATIONS

Each commercial FM broadcast station in the 88- to 108-MHz band is allocated a 150-kHz channel plus a 25-kHz guard band at both the upper and the lower edges of the station allocation by the FCC. Thus, a total channel width of 200 kHz is provided to each station in the commercial FM broadcast band.

$$150\,\text{kHz} + 2(25\,\text{kHz}) = 200\,\text{kHz}$$

In addition to this large bandwidth and guard-band combination (200 kHz), only alternate channels are assigned within any particular geographic area. In the UHF band, of which the commercial FM broadcast band is a part, reception is limited to distances only slightly farther than the horizon. Thus, assigning only alternate channels in any given geographic area limits the possibility of interference (see Fig. 4-4).

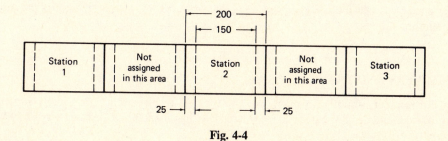

Fig. 4-4

*Mischa Schwartz, *Information Transmission, Modulation, and Noise*, McGraw-Hill, New York, 1959.

4.4 DEVIATION RATIO

The worst-case modulation index, in which the maximum permitted frequency deviation and the maximum permitted audio frequency are used, is called the *deviation ratio*.

$$\text{Deviation ratio} = \frac{\Delta f_{max}}{f_{a,\,max}} \tag{4.8}$$

Thus the deviation ratio for stations in the commercial FM broadcast band is

$$\text{Deviation ratio, 88–108 MHz} = \frac{75\,\text{kHz}}{15\,\text{kHz}} = 5$$

and for the sound portion of commercial television,

$$\text{Deviation ratio, TV} = \frac{25\,\text{kHz}}{15\,\text{kHz}} = 1.67$$

4.5 NARROWBAND FM VERSUS WIDEBAND FM

An examination of the Schwartz bandwidth curve of Fig. 4-3 indicates that at high values of m_f the curve tends towards a horizontal asymptote and at low values of m_f it tends toward the vertical. Detailed mathematical study would indicate that the bandwidth of an FM signal for which m_f is less than $\pi/2$ is dependent mainly upon the frequency of the modulating signal and is quite independent of frequency deviation. Further analysis would show that the bandwidth of an FM signal for which m_f is less than $\pi/2$ is equal to twice the modulating frequency.

$$\text{Bandwidth} = 2f_a \qquad \text{for } m_f < \pi/2$$

Just as with AM, and unlike the situation in which $m_f > \pi/2$, two side frequencies show up for each modulating frequency, one above and one below the frequency of the carrier, each spaced f_a away from the carrier frequency. Because of the limited bandwidth of FM signals with $m_f < \pi/2$, such modulations are referred to as *narrowband FM*, and FM signals with $m_f > \pi/2$ are referred to as *wideband FM*.

Though the spectrum for an AM signal and a narrowband FM signal appear to be the same, a Fourier analysis shows that the magnitude and phase relationships for AM and FM are quite different. See Fig. 4-5 for the frequency spectrum of a narrowband FM signal.

Many of the advantages obtained with wideband FM, such as noise reduction, are not available with narrowband FM. Why, then, would one want to use narrowband FM rather than AM? One reason is that with narrowband FM (as well as with wideband FM) the power content at the carrier frequency decreases as the modulation increases so that we have the desirable situation of putting the power where the information is.

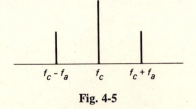

Fig. 4-5

4.6 FM RECEIVERS AND TRANSMITTERS

The FM receiver is similar in many ways to the AM receiver. Both are usually superheterodyne receivers. The commercial FM broadcast receiver usually has an intermediate frequency of 10.7 MHz. See Fig. 4-6.

Of course the demodulation circuit for FM receivers will be quite different from that used in an AM receiver. Other differences between the AM and FM receivers are the inclusion of a block called a *limiter* and one called a *de-emphasis network* in the FM receiver.

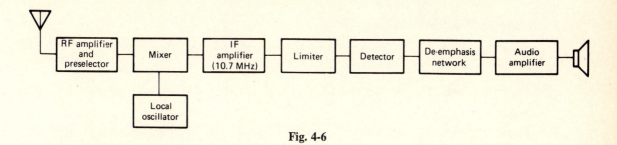

Fig. 4-6

The Limiter

The purpose of the limiter circuit is to clip all amplitude variations which may exist in the signal as it reaches this part of the system. This clipping removes any AM noise which may have become part of the signal. Clipping by the limiter eliminates noise but does not affect the information content of the signal because the information is contained in the frequency variations, not in the amplitude variations.

The De-emphasis Network

The de-emphasis network which appears in the block diagram of the FM receiver is only one-half of a system which consists of both a pre-emphasis and a de-emphasis network, the pre-emphasis network being located in the transmitter. The pre-emphasis network causes the higher-frequency information content of the audio signal at the transmitter to be amplified more than the lower-frequency information. The de-emphasis network compensates for this by reducing the gain of the higher-frequency audio signal. The reason for the inclusion of such a system is to reduce frequency-modulated noise which enters the transmitted signal while en route from the transmitter to the receiver as well as any such noise which may enter at the front end of the receiver.

Investigators found that noise which entered the signal as a frequency modulation occurred with greater likelihood and disturbance in the higher audio frequencies; thus, the pre-emphasis–de-emphasis system functions to reduce frequency-modulated noise.

FM Transmitters

As is most likely expected, the block diagram for an FM transmitter appears to be somewhat similar to the block diagram of an AM transmitter. Note in Fig. 4-7, the block diagram for an FM transmitter, the pre-emphasis network as expected from the discussion of FM receivers and a block entitled Exciter. The exciter is that portion of the FM transmitter within which modulation occurs.

There are two categories of techniques for the generation of an FM signal. One is called the *Direct method* and the other is called the *Indirect method.*

In the Direct method, a tuned circuit containing a device whose capacitance can be made to vary directly with the amplitude of the modulating signal is used. It is placed in shunt with a parallel RLC tank circuit. The most commonly used devices of this sort include the *transistor reactance modulator*, the *reactance tube modulator*, and *varactor diodes* (varicaps).

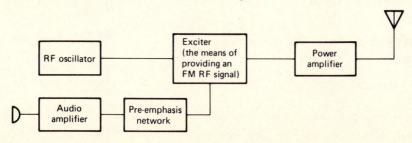

Fig. 4-7

The Transistor Reactance Modulator

Figure 4-8 is a schematic diagram of a transistor reactance modulator. The capacitance presented by this circuit is

$$C_{eq} = \frac{h_{fe} R_2 C_2}{h_{ie} + R_2} \tag{4.9}$$

The beta (β) of the transistor, h_{fe}, is caused to vary by changing the operating point of the transistor, the operating point being determined by the slowly varying audio signal input.

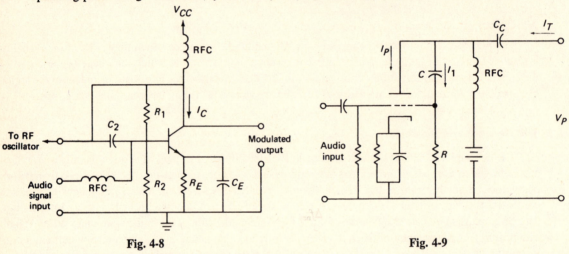

Fig. 4-8   Fig. 4-9

The Reactance Tube Modulator

Figure 4-9 is a schematic diagram of a reactance tube modulator. The capacitance presented by this circuit is

$$C_{eq} = g_m RC \tag{4.10}$$

where g_m is caused to vary with audio signal. A remote cutoff tube is usually used because the g_m of this tube is very sensitive to the operating point of the tube.

The Varactor Diode Modulator

Figure 4-10 is a schematic of a circuit which can function as an FM modulator. The varactor diode capacitance varies with its bias, which is determined by V_{CC} and the audio input signal.

The Indirect Method of FM Modulation

One of the difficulties encountered in FM transmitters which depend upon the Direct method of frequency modulation is that because of the variable nature of the tuning of the tank circuit, crystal-controlled oscillators cannot be used and therefore the stability inherent in such crystal-controlled units is not available.

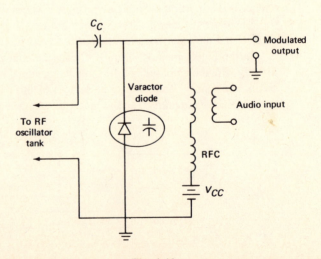

Fig. 4-10

An alternative technique for the generation of a frequency-modulated signal which permits the use of crystal control is called the Indirect method. In this technique the phase angle is caused to vary while holding the frequency constant. What is really generated by this technique is what is called a *phase-modulated signal*. With some minor doctoring, this phase-modulated signal can be passed off as an FM signal, and it is.

Frequency Multipliers

Most often, FM signals are initially generated in low-power circuits and circuits providing frequency deviations which are too small to meet FCC requirements.

The mathematical description of a frequency-modulated signal is

$$v = A \sin\left[2\pi(f_c + \Delta f \sin(2\pi f_a t))t\right] \tag{4.11}$$

The frequency is

$$f = f_c + \Delta f \sin 2\pi f_a t \tag{4.12}$$

Any means which will multiply the frequency of the FM signal by S will produce a new signal having a frequency deviation of $S\Delta f$.

$$Sf = S(f_c + \Delta f \sin 2\pi f_a t)$$
$$Sf = Sf_c + S\Delta f \sin 2\pi f_a t$$

Thus,

$$\Delta f_{\text{new}} = S\Delta f_{\text{old}} \tag{4.13}$$

with a new center frequency of $Sf_{c\,\text{old}}$.

Frequency multiplication is not difficult to obtain since harmonics which are generated by nonlinear devices such as class C amplifiers and varactor diodes provide outputs rich in harmonics, harmonics being signals having a frequency which is an integer multiple of the fundamental which is the input signal. It then becomes merely a chore of choosing the appropriate harmonic by using a frequency-selective circuit.

Frequency multipliers are generally limited by practical considerations to multiplications by 2, 3, or 4. Larger multiplication factors may be obtained by cascading these smaller multipliers.

Heterodyning

It sometimes becomes necessary to be able to adjust the frequency of the modulated signal without affecting frequency deviation. This can be accomplished by mixing, beating, or heterodyning, all three terms meaning the same thing. This is the same process used in the superheterodyne receiver to generate the intermediate frequency by heterodyning the local oscillator signal with the received signal.

The difference between heterodyning and multiplying is that in heterodyning the sinusoid angle is *added to* or *subtracted from*, while a multiplier *multiplies* the sinusoid angle by some factor. It is not unusual to find frequency multipliers followed by a heterodyner in an FM transmitter.

Solved Problems

4.1 A 107.6-MHz carrier is frequency modulated by a 7-kHz sine wave. The resultant FM signal has a frequency deviation of 50 kHz.

(*a*) Find the carrier swing of the FM signal.

(*b*) Determine the highest and lowest frequencies attained by the modulated signal.

(*c*) What is the modulation index of the FM wave?

SOLUTION

Given: $f_c = 107.6\,\text{MHz}$

               $f_a = 7\,\text{kHz}$

               $\Delta f = 50\,\text{kHz}$

Find: (a)   c.s.   (b) f_H, f_L (c) m_f

(a)   Relating carrier swing to frequency deviation

$$\text{c.s.} = 2\Delta f$$
$$= 2 \times 50 \times 10^3$$

$$\boxed{\text{c.s.} = 100\,\text{kHz}}$$

(b)   The upper frequency reached is equal to the rest or carrier frequency plus the frequency deviation:

$$f_H = f_c + \Delta f$$
$$= 107.6 \times 10^6 + 50 \times 10^3$$
$$= (107\,600 \times 10^3) + (50 \times 10^3)$$
$$= 107\,650 \times 10^3$$

$$\boxed{f_H = 107.65\,\text{MHz}}$$

The lowest frequency reached by the modulated wave is equal to the rest or carrier frequency minus the frequency deviation.

$$f_L = f_c - \Delta f$$
$$= 107.6 \times 10^6 - 50 \times 10^3$$
$$= 107\,600 \times 10^3 - 50 \times 10^3$$
$$= 107\,550 \times 10^3$$

$$\boxed{f_L = 107.55\,\text{MHz}}$$

(c)   The modulation index is determined by

$$m_f = \frac{\Delta f}{f_a}$$
$$= \frac{50 \times 10^3}{7 \times 10^3}$$

$$\boxed{m_f = 7.143}$$

4.2   Determine the frequency deviation and carrier swing for a frequency-modulated signal which has a resting frequency of 105.000 MHz and whose upper frequency is 105.007 MHz when modulated by a particular wave. Find the lowest frequency reached by the FM wave.

SOLUTION

Given: $f_0 = 105.000\,\text{MHz}$

               $f_{\text{upper}} = 105.007\,\text{MHz}$

Find: Δf,   c.s., f_{lower}

Frequency deviation is defined as the maximum change in frequency of the modulated signal away from the rest or carrier frequency.

$$\Delta f = (105.007 - 105.000) \times 10^6$$
$$= 0.007 \times 10^6$$
$$= 7000$$

$$\boxed{\Delta f = 7\,\text{kHz}}$$

Carrier swing can now be determined by

$$\text{c.s.} = 2\Delta f$$
$$= 2(7 \times 10^3)$$
$$= 14 \times 10^3$$

$$\boxed{\text{c.s.} = 14\,\text{kHz}}$$

The lowest frequency reached by the modulated wave can be found by subtracting the frequency deviation from the carrier or rest frequency.

$$f_{\text{lower}} = f_0 - \Delta_f$$
$$= (105.000 - 0.007) \times 10^6$$

$$\boxed{f_{\text{lower}} = 104.993\,\text{MHz}}$$

4.3   What is the modulation index of an FM signal having a carrier swing of 100 kHz when the modulating signal has a frequency of 8 kHz?

SOLUTION

Given: c.s. = 100 kHz

$f_a = 8$ kHz

Find: m_f

From the defining equation,

$$m_f = \frac{\Delta f}{f_a}$$

First determining Δf,

$$\Delta f = \frac{\text{c.s.}}{2}$$
$$= \frac{100 \times 10^3}{2}$$
$$= 50\,\text{kHz}$$

Now substituting into the equation for m_f,

$$m_f = \frac{50 \times 10^3}{8 \times 10^3}$$

$$\boxed{m_f = 6.25}$$

4.4   A frequency-modulated signal which is modulated by a 3-kHz sine wave reaches a maximum frequency of 100.02 MHz and minimum frequency of 99.98 MHz.

(a)   Determine the carrier swing.

(b)   Find the carrier frequency.

(c)   Calculate the frequency deviation of the signal.

(d)   What is the modulation index of the signal?

SOLUTION

Given: $f_{max} = 100.02$ MHz

$f_{min} = 99.98$ MHz

$f_a = 3$ kHz

Find: (a)   c.s.   (b) f_c (c) Δf (d) m_f

(a)   The carrier swing is defined as the total variation in frequency from the highest to lowest reached by the modulated wave.

$$\text{c.s.} = f_{max} - f_{min}$$
$$= 100.02 \times 10^6 - 99.98 \times 10^6$$
$$= 0.04 \times 10^6$$
$$= 40 \times 10^3$$

$$\boxed{\text{c.s.} = 40 \text{ kHz}}$$

(b)   The carrier frequency or rest frequency is midway between the maximum frequency and minimum frequency reached by the modulated wave.

$$f_c = \frac{f_{max} + f_{min}}{2}$$
$$= \frac{100.02 \times 10^6 + 99.98 \times 10^6}{2}$$
$$= 100 \times 10^6$$

$$\boxed{f_c = 100.00 \text{ MHz}}$$

(c)   Since the carrier swing is equal to twice the frequency deviation,

$$\Delta f = \frac{\text{c.s.}}{2}$$
$$= \frac{40 \times 10^3}{2}$$

$$\boxed{\Delta f = 20 \text{ kHz}}$$

(d)   The modulation index for a frequency modulated wave is defined as

$$m_f = \frac{\Delta f}{f_a}$$
$$= \frac{20 \times 10^3}{3 \times 10^3}$$

$$\boxed{m_f = 6.667}$$

4.5 An FM transmission has a frequency deviation of 20 kHz.

(*a*) Determine the percent modulation of this signal if it is broadcast in the 88–108 MHz band.

(*b*) Calculate the percent modulation if this signal were broadcast as the audio portion of a television broadcast.

SOLUTION

Given: $\Delta f = 20$ kHz

Find: (*a*) Percent modulation—FM broadcast band

        (*b*) Percent modulation—TV

(*a*) Percent modulation for an FM wave is defined as

$$M = \frac{\Delta f_{\text{actual}}}{\Delta f_{\text{max}}} \times 100$$

The maximum frequency deviation in the FM broadcast band permitted by the FCC is 75 kHz:

$$M = \frac{20 \times 10^3}{75 \times 10^3} \times 100$$

$$\boxed{M = 26.67\%}$$

(*b*)
$$M = \frac{\Delta f_{\text{actual}}}{\Delta f_{\text{max}}} \times 100$$

The maximum frequency deviation for the FM audio portion of a TV broadcast is 25 kHz as set by the FCC.

$$M = \frac{20 \times 10^3}{25 \times 10^3} \times 100$$

$$\boxed{M = 80.0\%}$$

4.6 (*a*) What is the frequency deviation and carrier swing necessary to provide 75% modulation in the FM broadcast band?

(*b*) Repeat for an FM signal serving as the audio portion of a TV broadcast.

SOLUTION

Given: $M = 75\%$

Find: (*a*) Δf_{FM}, c.s.$_{\text{FM}}$ (*b*) Δf_{TV}, c.s.$_{\text{TV}}$

(*a*) Frequency deviation is defined as

$$M = \frac{\Delta f_{\text{actual}}}{\Delta f_{\text{max}}} \times 100$$

The maximum frequency deviation permitted in the FM broadcast band, 88–108 MHz, by the FCC is 75 kHz.

$$75 = \frac{\Delta f_{\text{FM}}}{75 \times 10^3} \times 100$$

$$\Delta f_{\text{FM}} = \frac{75 \times 75 \times 10^3}{100}$$

$$= 56.25 \times 10^3$$

$$\boxed{\Delta f_{\text{FM}} = 56.25 \text{ kHz}}$$

Carrier swing is related to frequency deviation by

$$\text{c.s.}_{\text{FM}} = 2\Delta f_{\text{FM}}$$
$$= 2 \times 56.25 \times 10^3$$

$$\boxed{\text{c.s.}_{\text{FM}} = 112.5 \text{ kHz}}$$

(b)
$$M = \frac{\Delta f_{\text{actual}}}{\Delta f_{\text{max}}} \times 100$$

The maximum frequency deviation permitted by the FCC for the audio portion of a TV signal is 25 kHz. Thus,

$$75 = \frac{\Delta f_{\text{TV}}}{25 \times 10^3} \times 100$$
$$\Delta f_{\text{TV}} = \frac{75 \times 25 \times 10^3}{100}$$

$$\boxed{\Delta f_{\text{TV}} = 18.75 \text{ kHz}}$$

$$\text{c.s.}_{\text{TV}} = 2\Delta f_{\text{TV}}$$
$$= 2 \times 18.75 \times 10^3$$

$$\boxed{\text{c.s.}_{\text{TV}} = 37.5 \text{ kHz}}$$

4.7   Determine the percent modulation of an FM signal which is being broadcast in the 88–108 MHz band, having a carrier swing of 125 kHz.

SOLUTION

Given: c.s. = 125 kHz

Find: M

Frequency deviation and carrier swing are related by

$$\Delta f = \frac{\text{c.s.}}{2}$$
$$= \frac{125 \times 10^3}{2}$$
$$= 62.5 \text{ kHz}$$
$$M = \frac{\Delta f_{\text{actual}}}{\Delta f_{\text{max}}} \times 100$$

Maximum frequency deviation for the FM broadcast band permitted by the FCC is 75 kHz:

$$M = \frac{62.5 \times 10^3}{75 \times 10^3} \times 100$$

$$\boxed{M = 83.3\%}$$

4.8   The percent modulation of the sound portion of a TV signal is 80%. Determine the frequency deviation and carrier swing of the signal.

SOLUTION

Given: $M = 80\%$

Find: Δf,   c.s.

The percent modulation of an FM signal is

$$M = \frac{\Delta f_{\text{actual}}}{\Delta f_{\text{max}}} \times 100$$

The maximum frequency deviation for the sound portion of a TV signal as specified by the FCC is 25 kHz. Thus,

$$80 = \frac{\Delta f_{\text{actual}}}{25 \times 10^3} \times 100$$

$$\Delta f_{\text{actual}} = \frac{80 \times 25 \times 10^3}{100}$$

$$\boxed{\Delta f_{\text{actual}} = 20\,\text{kHz}}$$

Carrier swing is related to frequency deviation by

$$\text{c.s.} = 2\Delta f_{\text{actual}}$$
$$= 2 \times 20 \times 10^3$$

$$\boxed{\text{c.s.} = 40\,\text{kHz}}$$

4.9   A 5-kHz audio tone is used to modulate a 50-MHz carrier causing a frequency deviation of 20 kHz. Determine (a) the modulation index and (b) the bandwidth of the FM signal.

SOLUTION

Given: $f_a = 5\,\text{kHz}$

                  $f_c = 50.0\,\text{MHz}$

                  $\Delta f = 20\,\text{kHz}$

Find: (a) m_f (b) BW

(a) Modulation index is defined as

$$m_f = \frac{\Delta f}{f_a}$$
$$= \frac{20 \times 10^3}{5 \times 10^3}$$

$$\boxed{m_f = 4}$$

(b) Referring to the Schwartz bandwidth curve, Fig. 4-3, and entering on the horizontal axis with $m_f = 4$, it is found that

$$\frac{\text{BW}}{\Delta f} = 3.8$$

This is shown in Fig. 4-11.
Substituting 20×10^3 for Δf as given,

$$\frac{\text{BW}}{20 \times 10^3} = 3.8$$

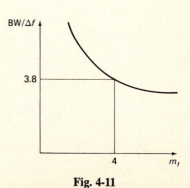

Fig. 4-11

Solving for BW,

$$BW = 3.8 \times 20 \times 10^3$$
$$= 76 \times 10^3$$

$$\boxed{BW = 76\,kHz}$$

4.10   Determine the frequency of the modulating signal which is producing an FM signal having a bandwidth of 50 kHz when the frequency deviation of the FM signal is 10 kHz.

SOLUTION

Given: BW = 50 kHz

$\Delta f = 10$ kHz

Find: f_a

In order to find f_a, reference must be made to the Schwartz bandwidth curve, Fig. 4-3. In order to enter this curve, determine BW/Δf:

$$\frac{BW}{\Delta f} = \frac{50 \times 10^3}{10 \times 10^3}$$
$$= 5$$

From Fig. 4-3,

$$m_f = 2$$
$$= \frac{\Delta f}{f_a}$$

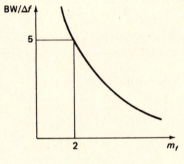

Fig. 4-12

So,

$$2 = \frac{10 \times 10^3}{f_a}$$
$$f_a = \frac{10 \times 10^3}{2}$$

$$\boxed{f_a = 5\,kHz}$$

This is shown in Fig. 4-12.

4.11   A 103.0-MHz carrier is frequency modulated by a 10-kHz sine wave. Determine the modulation index of the FM signal. Referring to Schwartz's curve, Fig. 4-3, determine the bandwidth when the carrier swing is 80 kHz.

SOLUTION

Given: $f_c = 103.0$ MHz

$f_a = 10$ kHz

c.s. = 80 kHz

Find: m_f,   BW

The defining equation for the modulation index is

$$m_f = \frac{\Delta f}{f_a}$$

However, before using this equation, it is necessary to determine the frequency deviation, Δf.

$$\Delta f = \frac{c.s.}{2}$$

$$= \frac{80\,kHz}{2}$$

$$= 40\,kHz$$

Returning to the defining equation for modulation index:

$$m_f = \frac{40 \times 10^3}{10 \times 10^3}$$

$$\boxed{m_f = 4}$$

Entering the Schwartz bandwidth curve of Fig. 4.3 with $m_f = 4$ results in

$$\frac{BW}{\Delta f} = 3.5$$

$$BW = 3.5 \times 40 \times 10^3$$

$$= 140 \times 10^3$$

$$\boxed{BW = 140\,kHz}$$

4.12 If a 6-MHz band were being considered for use with the same standards that apply to the 88–108 MHz band, how many FM stations could be accommodated?

SOLUTION

Given: BW = 6 MHz

Find: Number of stations

Each station requires a total bandwidth of 400 kHz; 150 kHz for the signal and a 25-kHz guard band above and below with only alternate channels used.

$$\text{Number of stations} = \frac{6 \times 10^6}{400 \times 10^3}$$

$$\boxed{\text{Number of stations} = 15}$$

4.13 Determine the bandwidth of a narrowband FM signal which is generated by a 4-kHz audio signal modulating a 125-MHz carrier.

SOLUTION

Given: Narrowband FM

$\quad\quad\quad\quad f_a = 4\,kHz$

$\quad\quad\quad\quad f_c = 125\,MHz$

Find: BW

Since this is a narrowband FM signal, the bandwidth is found merely by doubling the modulating frequency:

$$BW = 2f_a$$

$$= 2 \times 4 \times 10^3$$

$$\boxed{BW = 8\,kHz}$$

4.14 A 2-kHz audio signal modulates a 50-MHz carrier, causing a frequency deviation of 2.5 kHz. Determine the bandwidth of the FM signal.

SOLUTION

Given: $f_a = 2\,\text{kHz}$

$f_c = 50\,\text{MHz}$

$\Delta f = 2.5\,\text{kHz}$

Find: BW

$$m_f = \frac{\Delta f}{f_a} = \frac{2.5 \times 10^3}{2 \times 10^3}$$

$$= 1.25$$

Since this is less than $\pi/2$, we are dealing with a narrowband signal; thus,

$$\text{BW} = 2f_a$$

$$\boxed{\text{BW} = 2 \times 2 \times 10^3 = 4\,\text{kHz}}$$

4.15 The transistor reactance modulator shown in Fig. 4-13 is to be used in an FM transmitter. The input resistance of the transistor h_{ie}, is 600 Ω and the beta of the transistor is 65.

(a) How much capacitance does this circuit present to the tank circuit it is attached to?

(b) If the beta of the transistor can be caused to swing from 50 to 75, determine the swing in capacitance presented by this circuit.

SOLUTION

Given: $h_{ie} = 600\,\Omega$

$\beta_0 = 65$

$\beta_1 = 50$

$\beta_2 = 75$

Find: (a) C_{eq2} (b) C_{eq1}, C_{eq2}

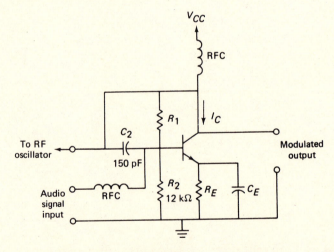

Fig. 4-13

(a) The equation for determining the equivalent capacitance of a transistor reactance modulator is

$$C_{eq} = \frac{h_{fe} R_2 C_2}{h_{ie} + R_2}, \qquad \text{where} \qquad h_{fe} = \beta$$

Substituting (values for C_2 and R_2 are taken from Fig. 4-13),

$$C_{eq} = \frac{65(12 \times 10^3)(150 \times 10^{-12})}{600 + 12\,000}$$

$$= 9.29 \times 10^{-9}$$

$$\boxed{\begin{array}{c} C_{eq} = 9.29\,\text{nF} \\ \text{or} \\ 9290\,\text{pF} \end{array}}$$

(b)   Using the equation for equivalent capacitance of the transistor reactance modulator,

$$C_{eq1} = \frac{h_{fe1} R_2 C_2}{h_{ie} + R_2}$$

and substituting numerical values,

$$C_{eq1} = \frac{(50)(12 \times 10^3)(150 \times 10^{-12})}{600 + 12\,000}$$

$$= 7.143 \times 10^{-9}$$

The lower value of equivalent capacitance reached is therefore

$$\boxed{\begin{array}{c} C_{eq1} = 7.14\,\text{nF} \\ \text{or} \\ 7140\,\text{pF} \end{array}}$$

Again using

$$C_{eq2} = \frac{h_{fe2} R_2 C_2}{h_{ie} + R_2}$$

and substituting appropriate numerical values,

$$C_{eq2} = \frac{(75)(12 \times 10^3)(150 \times 10^{-12})}{600 + 12\,000}$$

Thus the higher value reached is

$$C_{eq2} = 10.71 \times 10^{-9}$$

So,

$$\boxed{\begin{array}{c} C_{eq2} = 10.71\,\text{nF} \\ \text{or} \\ = 10\,710\,\text{pF} \\ \text{or} \\ = 0.010\,71\,\mu\text{F} \end{array}}$$

4.16   The reactance tube modulator shown in Fig. 4-14 uses a remote cutoff tube whose transconductance g_m varies from 2500 μS to 3500 μS (in the SI metric system, the unit of conductance is the siemens, abbreviated S). Determine the range of capacitance it presents.

SOLUTION

Given: $g_{m1} = 2500\ \mu$S

$g_{m2} = 3500\ \mu$S

$C = 75$ pF (from Fig. 4-14)

$R = 100$ kΩ (from Fig. 4-14)

Find: C_{eq1}, C_{eq2}

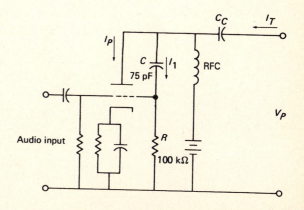

Fig. 4-14

The appropriate formula for the equivalent capacitance of a reactance tube modulator is

$$C_{eq} = g_m RC$$

Using the lower value of g_m,

$$C_{eq1} = (2500 \times 10^{-6})(100 \times 10^3)(75 \times 10^{-12})$$
$$= 18.75 \times 10^{-9}$$

Thus the lower value reached by this reactance tube modulator is

$$C_{eq1} = 18.75\,\text{nF}$$
$$\text{or}$$
$$= 18\,750\,\text{pF}$$
$$\text{or}$$
$$= 0.018\,75\,\mu\text{F}$$

Using the same formula to determine the high value reached by C_{eq},

$$C_{eq2} = g_{m2} RC$$
$$= (3500 \times 10^{-6})(100 \times 10^3)(75 \times 10^{-12})$$
$$= 26.25 \times 10^{-9}$$

Thus the highest value reached by C_{eq} is

$$C_{eq2} = 26.25\,\text{nF}$$
$$\text{or}$$
$$= 26\,250\,\text{pF}$$
$$\text{or}$$
$$= 0.026\,25\,\mu\text{F}$$

4.17 Figure 4-15 is the block diagram of the frequency multiplier and heterodyne portion of an FM transmitter. Calculate the carrier frequency and frequency deviation of each of the points: (a) 1, (b) 2, and (c) 3.

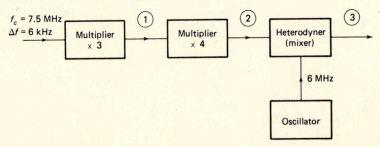

Fig. 4-15

SOLUTION

Given: $f_{c0} = 7.5\,\text{MHz}$
         $\Delta f_0 = 6\,\text{kHz}$
         $S_1 = 3$
         $S_2 = 4$
         $f_{osc} = 6\,\text{MHz}$

Find: (a) $f_{c1}, \Delta f_1$ (b) $f_{c2}, \Delta f_2$ (c) $f_{c3}, \Delta f_3$

(a) The effect of the first multiplier is to multiply both carrier frequency and frequency deviation by 3.

$$f_{c1} = S_1 f_{c0}$$
$$= 3(7.5 \times 10^6)$$
$$= 22.5 \times 10^6$$

$$\boxed{f_{c1} = 22.5\,\text{MHz}}$$

$$\Delta f_1 = S \, \Delta f_0$$
$$= 3(6 \times 10^3)$$
$$= 18 \times 10^3$$

$$\boxed{\Delta f_1 = 18\,\text{kHz}}$$

(b) The effect of the second multiplier is to multiply the carrier frequency and frequency deviation present at its input by 4.

$$f_{c2} = S_2 f_{c1}$$
$$= 4(22.5 \times 10^6)$$

$$\boxed{f_{c2} = 90\,\text{MHz}}$$

$$\Delta f_2 = S_2 \, \Delta f_1$$
$$= 4(18 \times 10^3)$$
$$= 72 \times 10^3$$

$$\boxed{\Delta f_2 = 72\,\text{kHz}}$$

(c) The mixer has the effect of raising or lowering the carrier frequency by an amount equal to the oscillator frequency. Assuming a rise in frequency,

$$f_{c3} = f_{c2} + f_{osc}$$
$$= (90 \times 10^6) + (6 \times 10^6)$$
$$= 96 \times 10^6$$

$$\boxed{f_{c3} = 96\,\text{MHz}}$$

The heterodyning caused by the mixer and oscillator does not change the frequency deviation. Therefore

$$\Delta f_3 = \Delta f_2$$
$$\boxed{\Delta f_3 = 72\,\text{kHz}}$$

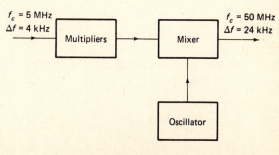

4.18 A 50-MHz FM signal is desired which is to have a frequency deviation of 24 kHz. The output of early stages of the transmitter is a 5-MHz signal with a frequency deviation of 4 kHz. How can the desired output be obtained? See Fig. 4-16.

Fig. 4-16

SOLUTION

Given: $f_{c\,out} = 50\text{ MHz}$

$\Delta f_{out} = 24\text{ kHz}$

$f_{c\,in} = 5\text{ MHz}$

$\Delta f_{in} = 4\text{ kHz}$

Find: System

First, find the amount of frequency multiplication required in order to provide the desired frequency deviation.

$$\Delta f_{out} = S\,\Delta f_{in}$$
$$24 \times 10^3 = S(4 \times 10^3)$$
$$S = \frac{24 \times 10^3}{4 \times 10^3}$$

$$\boxed{S = 6}$$

Since it is most convenient to specify frequency multipliers which multiply by 2, 3, or 4, a cascade arrangement of a $2 \times$ multiplier and a $3 \times$ multiplier is required.

This frequency multiplication, although it provides an appropriate value of frequency deviation, doesn't necessarily cause the carrier frequency to have the required value. The carrier frequency obtained by passing the signal through the multipliers is found from

$$f_{c2} = Sf_{c1}$$
$$= 6(5 \times 10^6)$$
$$= 30 \times 10^6$$

$$\boxed{f_{c2} = 30\text{ MHz}}$$

This frequency of carrier can be caused to change without affecting the frequency deviation by passing the signal through a heterodyne section made up of a mixer and oscillator. The frequency of the oscillator can be found as follows:

$$f_{c2} + f_{osc} = f_{c3}$$
$$30 \times 10^6 + f_{osc} = 50 \times 10^6$$
$$f_{osc} = (50 \times 10^6) - (30 \times 10^6)$$
$$= 20 \times 10^6$$

$$\boxed{f_{osc} = 20\text{ MHz}}$$

Supplementary Problems

4.19 A 93.2-MHz carrier is frequency modulated by a 5-kHz sine wave. The resultant FM signal has a frequency deviation of 40 kHz.

(a)   Find the carrier swing of the FM signal.

(b)   Determine the highest and lowest frequencies attained by the modulated signal.

(c)   What is the modulation index of the FM wave?

Ans.   (a) 80 kHz, (b) 93.16 MHz, 93.24 MHz, (c) 8

4.20   Determine the carrier swing, the highest and lowest frequencies attained, and the modulation index of the FM signal generated by frequency modulating a 101.6-MHz carrier with an 8-kHz sine wave causing a frequency deviation of 40 kHz.   *Ans.*   80 kHz; 101.64 MHz, 101.56 MHz; 5

4.21   Calculate the frequency deviation and carrier swing of a frequency-modulated wave which was produced by modulating a 50.400 MHz carrier. The highest frequency reached by the FM wave is 50.406 MHz. Then calculate the lowest frequency reached by the FM wave.   *Ans.*   6 kHz, 12 kHz, 50.394 MHz

4.22   Find the upper and lower frequencies that are reached by a frequency-modulated wave that has a rest frequency of 104.003 MHz and a frequency deviation of 60 kHz. What is the carrier swing of the modulated signal?
Ans.   104.063 MHz, 103.943 MHz, 120 kHz

4.23   Determine the frequency deviation and carrier swing for a frequency-modulated signal which has a resting frequency of 97.340 MHz and whose upper frequency is 97.350 MHz when modulated by a particular audio sine wave. Find the lower frequency reached by the FM wave.   *Ans.*   10 kHz, 20 kHz, 97.330 MHz

4.24   The carrier swing of a frequency-modulated signal is 120 kHz. The modulating signal is a 6-kHz sine wave. Determine the modulation index of the FM signal.   *Ans.*   10

4.25   Determine the modulation index of a frequency-modulated signal having a frequency deviation of 70 kHz. The modulating signal has a frequency of 10 kHz.   *Ans.*   7

4.26   A 12-kHz sine wave is to frequency modulate a carrier causing a carrier swing of 80 kHz. Determine the modulation index.   *Ans.*   3.333

4.27   Determine the carrier swing, carrier frequency, frequency deviation, and modulation index for a frequency-modulated signal which reaches a maximum frequency of 99.047 MHz and a minimum frequency of 99.023 MHz. The frequency of the modulating signal is 7 kHz.   *Ans.*   24 kHz, 99.035 MHz, 12 kHz, 1.7

4.28   A carrier is frequency modulated by a 4-kHz sine wave resulting in an FM signal having a maximum frequency of 107.218 MHz and a minimum frequency of 107.196 MHz.

(*a*)   Find the carrier swing.

(*b*)   Calculate the carrier frequency.

(*c*)   What is the frequency deviation of the FM signal?

(*d*)   Determine the modulation index of the FM signal.

Ans.   (*a*) 22 kHz, (*b*) 107.207 MHz, (*c*) 11 kHz, (*d*) 2.75

4.29   A frequency-modulated signal reaches a maximum frequency of 64.073 MHz and a minimum frequency of 64.050 MHz when modulated by a 4.5-kHz sine wave. Determine the carrier swing, the carrier frequency, the frequency deviation, and the modulation index of the FM signal.
Ans.   23 kHz, 64.0615 MHz, 11.5 kHz, 2.56

4.30   An FM signal for broadcast in the 88–108 MHz band has a frequency deviation of 15 kHz. Find the percent modulation of this signal. If this signal were prepared for broadcast as the audio portion of a television program, what would the percent modulation be?   *Ans.*   20%, 60%

4.31   The audio portion of a TV broadcast is to be modulated 80%. Determine the resultant frequency deviation and carrier swing.   *Ans.*   20 kHz, 40 kHz

4.32   An FM signal to be broadcast in the 88–108 MHz FM broadcast band is to be modulated 70%. Determine the frequency deviation and carrier swing.   *Ans.*   52.5 kHz, 105 kHz

4.33   Calculate the frequency deviation and carrier swing necessary to provide an 80% modulation in the FM broadcast band. Repeat this for an FM signal serving as the audio portion of a TV broadcast.
Ans.   60 kHz, 120 kHz; 20 kHz, 40 kHz

4.34   Find the percent modulation of an FM signal which is being broadcast in the 88–108 MHz band having a carrier swing of 150 kHz.   *Ans.*   100%

4.35   Determine the frequency deviation and carrier swing of an FM signal which is the audio portion of a TV signal and has a percent modulation of 85%.   *Ans.*   21.25 kHz, 42.5 kHz

4.36   An 8-kHz audio tone is used to modulate a 50.0-MHz carrier causing a frequency deviation of 20 kHz. Determine the modulation index and the bandwidth of the FM signal.   *Ans.*   2.5, 100 kHz

4.37   Calculate the modulation index and bandwidth of an FM signal having a 50-kHz frequency deviation when modulated by a 7-kHz audio tone.   *Ans.*   7.14, 150 kHz

4.38   Find the modulation index and bandwidth of an FM signal generated by modulating a 100.0-MHz carrier with a 3.8-kHz audio tone causing a carrier swing of 60 kHz.   *Ans.*   7.89, 87 kHz

4.39   Determine the frequency of the modulating signal which is producing an FM signal having a bandwidth of 60 kHz when the frequency deviation of the FM signal is 20 kHz.   *Ans.*   2.857 kHz

4.40   An FM signal has a bandwidth of 100 kHz when its frequency deviation is 25 kHz. Find the frequency of the modulating signal.   *Ans.*   7.1 kHz

4.41   Find the frequency of the audio signal which is frequency modulating a 100-MHz carrier causing a frequency deviation of 25 kHz resulting in an FM signal having a bandwidth of 80 kHz.   *Ans.*   4.5 kHz

4.42   An FM signal has a frequency deviation of 40 kHz when the bandwidth of the signal is 160 kHz. Determine the frequency of the modulating signal.   *Ans.*   11.4 kHz

4.43   A 15-kHz sine wave is frequency modulating a 104.500-MHz carrier. Determine the modulation index of the FM signal and determine the bandwidth of the FM signal if the carrier swing is 130 kHz.   *Ans.*   214.5 kHz

4.44   If an 18-MHz band were to be considered for use with the same standards that apply to the 88–108 MHz FM broadcast band, how many FM stations could be accommodated?   *Ans.*   45

4.45   What is the bandwidth of a narrowband FM signal which is generated by a 5-kHz audio signal modulating a 115-MHz carrier?   *Ans.*   10 kHz

4.46   A 50.004-MHz carrier is to be frequency modulated by a 3-kHz audio tone resulting in a narrowband FM signal. Determine the bandwidth of the FM signal.   *Ans.*   6 kHz

4.47   Determine the bandwidth of a signal generated by a 2.8-kHz audio tone frequency modulating a 98.004-MHz carrier resulting in a frequency deviation of 3.00 kHz.   *Ans.*   5.6 kHz

4.48   A frequency deviation of 4 kHz results from frequency modulating a 106.00-MHz carrier. The modulating signal is a 3500-Hz sine wave. Determine the bandwidth of the FM signal.   *Ans.*   7000 Hz

4.49   What function is served by the limiter in an FM receiver?   *Ans.*   To remove amplitude variations.

4.50   Why does the limiter not affect the information content of the signal?
Ans.   The information is contained in the frequency variations.

4.51 How does the function of the pre-emphasis–de-emphasis network differ from the function of the limiter circuits?
Ans. Pre-emphasis–de-emphasis system removes FM noise. Limiter removes AM noise.

4.52 What is an exciter? *Ans.* The means of providing an FM signal.

4.53 How does the Direct method of frequency modulation differ from the Indirect method?
Ans. In the Indirect method, a phase-modulated signal is first generated.

4.54 What is a varicap? What is its function in the exciter section of an FM transmitter?
Ans. A capacitor whose capacitance varies with applied voltage. Used in variable oscillators.

4.55 The input resistance of the transistor shown as part of the transistor reactance modulator of Fig. 4-17 is 450 Ω while the beta of the transistor is 80. What value of capacitance does this circuit present to the tank it is placed in shunt with? If the beta of the transistor swings between 60 and 100, what are the lower and upper values of capacitance presented by this circuit? *Ans.* 39.11 nF; 29.33 nF, 48.89 nF

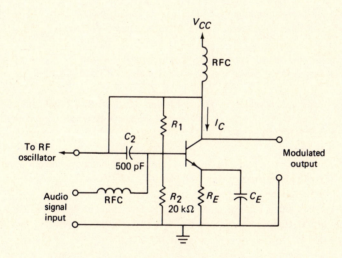

Fig. 4-17

4.56 Repeat Problem 4.55 if the circuit component values are changed to $C_2 = 400$ pF and $R_2 = 8$ kΩ. The same transistor is used. *Ans.* 30.28 nF; 22.71 nF, 37.86 nF

4.57 What new values of equivalent capacitance are presented by the circuit of Fig. 4-17 if the transistor is replaced with one whose beta is 90 and swings between 75 and 105? The input resistance of the replacement transistor is 1500 Ω. *Ans.* 41.85 nF, 34.88 nF, 48.83 nF

4.58 R_2 and C_2 of Fig. 4-17 are replaced with a resistor of 16 kΩ and a capacitor of 150 pF. The transistor to be used with this reactance modulator has a beta that swings between 35 and 60. The transistor has an input resistance of 1200 Ω. Calculate the lower and upper values of capacitance that are reached by this modulator.
Ans. 4.88 nF, 8.37 nF

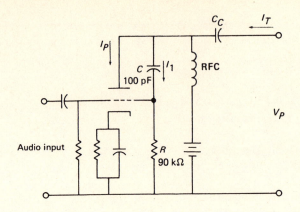

Fig. 4-18

4.59 The transconductance of the vacuum tube used in the reactance tube modulator of Fig. 4-18 is made to vary from 2800 μS to 4300 μS. Calculate the range of equivalent capacitance it presents.   *Ans.*   25.2 nF, 38.7 nF

4.60 Determine the swing in equivalent capacitance presented by the circuit of Fig. 4-18 if the tube used has a transconductance that ranges from 1500 μS to 1900 μS.   *Ans.*   13.5 nF, 17.1 nF

4.61 If the resistance and capacitance values of Fig. 4-18 were changed to 110 kΩ and 150 pF and the range of transconductance of the tube was from 3100 μS to 4000 μS, calculate the range of equivalent capacitance presented by this circuit.   *Ans.*   51.15 nF, 66 nF

4.62 Why can't crystal-controlled oscillators be used in the Direct method generation of frequency-modulated signals?   *Ans.*   The output frequency is not variable.

4.63 What is it about class C amplifiers and varactor diodes that permits their use as frequency multipliers?
Ans.   They are nonlinear devices.

4.64 Figure 4-19 is the block diagram of the frequency-multiplication and heterodyne section of an FM transmitter. Determine the carrier frequency and frequency deviation at points 1, 2, and 3.
Ans.   27 MHz, 17 kHz, 81 MHz, 51 kHz; 101 MHz; 51 kHz

4.65 Repeat Problem 4-64 if the input carrier frequency is 12 MHz, the input frequency deviation is 6 kHz, and the oscillator frequency is 18 MHz.   *Ans.*   24 MHz, 12 kHz; 72 MHz, 36 kHz; 90 MHz, 36 kHz

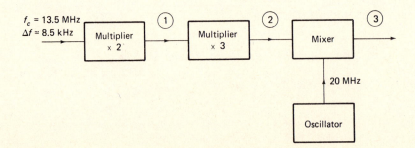

Fig. 4-19

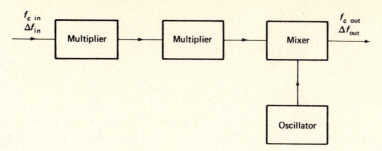

Fig. 4-20

4.66 In Fig. 4-20 determine the appropriate multiplier values and oscillator frequency so as to provide an output FM signal having a carrier frequency of 106 MHz with a frequency deviation of 60 kHz if the input is an FM signal having a carrier frequency of 9 MHz and a frequency deviation of 10 kHz.   *Ans.*   ×2, ×3, 52 MHz

4.67 The block diagram of Fig. 4-20 is required to produce an output FM signal having a carrier frequency of 54 MHz and a frequency deviation of 24 kHz when presented with an FM signal at its input having a carrier frequency of 7.0 MHz and a frequency deviation of 4 kHz. Determine the multiplier values and frequency the oscillator should be tuned to in order to meet these specifications.   *Ans.*   ×3, ×2, 12 MHz

4.68 What is the difference between frequency multiplication and heterodyning?
Ans.   When heterodyning, we are dealing with addition and subtraction.

Chapter 5

Transmission Lines

INTRODUCTION

Electromagnetic waves travel in free space at 3×10^8 meters/s or 186 000 miles/s. In other than free space, electromagnetic waves travel a bit slower; however, as a first approximation the speed in free space can be assumed to be the speed on a transmission line. If a signal varies so rapidly or the line is so long that before the leading edge of the signal reaches the end of the transmission line the incoming signal undergoes an appreciable change, we must consider how the transmission line affects the signal.

5.1 PULSE ON A TRANSMISSION LINE

Consider a pulse launched on a 100-meter transmission line. The pulse is to be of such short duration that the leading edge does not arrive at the load end of the line before the trailing edge of the pulse has left the generator. The amount of time for the signal to travel the length of the transmission line can be calculated as follows:

$$\frac{100 \text{ meters}}{3 \times 10^8 \text{ meters/s}} = 33.3 \times 10^{-8} \text{ s} \quad \text{or} \quad 333 \text{ ns}$$

If the pulse width is less than 333 ns in duration, the voltage at the generator will have returned to zero before the leading edge of the pulse arrives at the load. Figure 5-1 describes this situation.

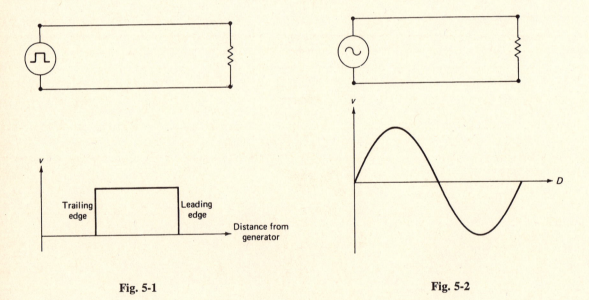

Fig. 5-1   Fig. 5-2

5.2 A SINE WAVE ON A TRANSMISSION LINE

Now consider a *sine wave* traveling on a transmission line whose length is appreciable compared to the wavelength of the sine wave. The wavelength of the sine wave is defined as the distance traveled by an electromagnetic signal during one cycle of the sine wave. Figure 5-2 shows a sine wave on a transmission line. The sine wave in this case has a wavelength equal to the physical length of the line.

Since the instantaneous value of a sine wave is constantly changing, we can imagine the ramifications involved when the length of the transmission line is of the same order of magnitude as the wavelength of the signal traveling on it. The instantaneous voltage anywhere on the line will be different from that elsewhere on the line.

The reason this situation has not been encountered with transmission lines which were much shorter than a wavelength is that although the voltage everywhere on the line was different, the difference was very minute since the speed of propagation was much greater than the rate of change of the signal. Consider the wavelength of a 60-Hz sine wave:

$$f\lambda = 3 \times 10^8 \text{ meters/s}$$
$$60\lambda = 3 \times 10^8 \text{ meters/s}$$
$$\lambda = \frac{3 \times 10^8}{60} = 0.05 \times 10^8$$

$$\boxed{\lambda = 5 \times 10^6 \text{ meters}}$$

or

$$f\lambda = 186\,000 \text{ miles/s}$$
$$60\lambda = 186\,000$$
$$\lambda = \frac{186\,000}{60}$$

$$\boxed{\lambda = 3100 \text{ miles}}$$

Thus we see that the wavelength of a 60-Hz signal is 5 million meters or 3100 miles. Unless the transmission line carrying the 60-Hz signal is an appreciable part of this distance, transmission-line theory need not be considered.

5.3 CHARACTERISTIC IMPEDANCE

Transmission lines whose lengths are an appreciable part or multiple of a wavelength of the signal being transmitted on it are described by a parameter referred to as characteristic impedance Z_0. The characteristic impedance is the impedance that a theoretically *infinite* length of this cable would present at its input end.

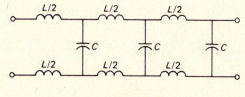

Fig. 5-3

Since every section of cable has a capacitance and an inductance, an infinite length of cable can be considered as an infinite network of inductors and capacitors. See Fig. 5-3.

The relationship between capacitance and inductance per unit length and characteristic impedance is $Z_0 = \sqrt{L/C}$.

The characteristic impedance of a number of common cables used as transmission lines is shown in Table 5-1.

Table 5-1

Type #	Description	Characteristic Impedance (Ω)
RG 8/U	Coaxial cable	52
TG11 A/U	Coaxial cable	75
214-056	Twin lead (commonly used for TV lead-in)	300
	Air-insulated parallel conductors with ceramic spacers	200–600

5.4 REFLECTED WAVES AND STANDING-WAVE RATIO (SWR)

Due to power and energy considerations at the load end of a short-circuited or open-circuited transmission line, an argument can be made for the existence of waves reflected from the load.

When a transmission line is terminated in any load other than a resistance having a magnitude equal to the characteristic impedance of the line, a reflected wave as well as an incident wave is present on the line. The summation of the incident and reflected wave at each point on the line gives rise to different rms voltage values at different points on the line.

A voltmeter placed at each point on the transmission line indicates an rms voltage at each point which varies from point to point on the line as shown in Fig. 5-4.

The ratio of the largest rms value to the smallest rms value of voltage on the line is called the *voltage standing-wave ratio* (VSWR). The largest rms and smallest rms values are measured at different points on the line separated by a distance equal to a quarter wavelength.

Similarly, measurement of the rms value of current at each point on the transmission line gives rise to different values at each point on the line as shown in Fig. 5-5.

The ratio of the smallest rms current value to the largest is called the *current standing-wave ratio* (ISWR).

The VSWR and the ISWR are equal. Frequently SWR is used in place of the terms VSWR and ISWR.

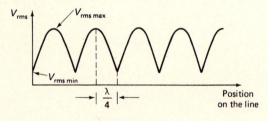

Fig. 5-4

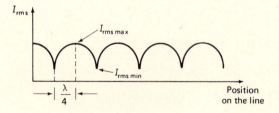

Fig. 5-5

5.5 STANDING-WAVE RATIO

The optimum condition for transmitting power to a load over a transmission line is one in which the maximum rms values of voltage and current are equal to the minimum rms values of voltage and current.

The SWR is an indication of how close or how far we are to the optimum condition for transmitting power to the load. The closer the SWR comes to $1:1$, the closer the best or optimum condition is realized. In determining the SWR, a ratio is set up with the larger quantity taken first:

$$\text{SWR} = V_{\text{rms max}} : V_{\text{rms min}} = I_{\text{rms max}} : I_{\text{rms min}}$$

Laboratory and field investigation shows that

$$\text{SWR} = Z_L : Z_0 \tag{5.1}$$

the SWR being a measure of the mismatch between load and line. As with all ratios, the SWR can also be represented as a fraction.

$$\text{SWR} = \frac{V_{\text{rms max}}}{V_{\text{rms min}}} = \frac{I_{\text{rms max}}}{I_{\text{rms min}}} \tag{5.2}$$

$$\text{SWR} = \frac{Z_L}{Z_0} \tag{5.3}$$

5.6   THE REFLECTION COEFFICIENT K_r

Another factor to be considered when dealing with mismatched load and transmission lines is the reflection coefficient K_r. K_r is defined as the reflected voltage divided by the incident voltage, or alternatively as reflected current divided by incident current.

Since in the optimum condition the reflected wave goes to zero, the optimum value for K_r is zero.

$$\boxed{K_r = \frac{V_{\text{refl}}}{V_{\text{inc}}}} \tag{5.4}$$

$$\boxed{K_r = \frac{I_{\text{refl}}}{I_{\text{inc}}}} \tag{5.5}$$

Since both SWR and K_r are indications of how poor a mismatch exists, there should be a relationship between them. Such a relationship does exist:

$$\boxed{\text{SWR} = \frac{K_r + 1}{1 - K_r}} \tag{5.6}$$

It should also be possible to express the reflection coefficient in terms of the load resistance and the characteristic impedance, their inequality to each other being the main cause of K_r being other than zero. This relationship also exists.

$$\boxed{K_r = \left| \frac{Z_L - Z_0}{Z_0 + Z_L} \right|} \tag{5.7}$$

5.7   REFLECTED POWER

Since the primary reason for launching a wave on a transmission line is to transfer power from a source to a load, we are concerned with how effectively power has been injected into the load. Since power is equal to the square of the voltage divided by the resistance of the load, and since the reflection coefficient is reflected voltage divided by incident voltage, we can write the following:

$$P = \frac{V^2}{R}$$

$$P_{\text{refl}} = \frac{V_{\text{refl}}^2}{R_L}$$

$$P_{\text{inc}} = \frac{V_{\text{inc}}^2}{R_L}$$

$$\frac{P_{\text{refl}}}{P_{\text{inc}}} = \frac{V_{\text{refl}}^2/R_L}{V_{\text{inc}}^2/R_L}$$

$$= \frac{V_{\text{refl}}^2}{V_{\text{inc}}^2}$$

$$K_r = \frac{V_{\text{refl}}}{V_{\text{inc}}}$$

$$\boxed{\frac{P_{\text{refl}}}{P_{\text{inc}}} = K_r^2}$$ (5.8)

5.8   VELOCITY FACTOR

Thus far in this chapter, the speed of electromagnetic waves on a transmission line has been approximated as being equal to the speed of electromagnetic waves in free space; 3×10^8 meters/s. This is not exactly true. In fact, the ratio of the speed of electromagnetic waves on a particular transmission line to that in free space is known as the velocity factor k. The velocity factor for most common transmission lines varies from a low of 0.55 for certain twisted pairs to 0.98 for small conductors widely spaced as open-wire lines.

5.9   QUARTER-WAVE MATCHING TRANSFORMERS

Quarter-wavelength sections of transmission line can be used to match a load to a transmission when the resistance of the load is not equal to the characteristic impedance of the transmission line. Since the quarter-wavelength section of line functions somewhat as a matching transformer, it is referred to as a quarter-wave matching transformer. The proper value of characteristic impedance for the quarter-wave transmission line section can be found from

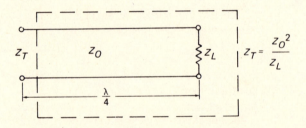

Fig. 5-6

$$Z_T = \frac{Z_0^2}{Z_L}$$ (5.9)

where Z_T is the impedance seen looking into the quarter-wave matching section when it is attached to the load Z_L, and Z_0 is the characteristic impedance of the quarter-wave matching section necessary to provide a match. See Fig. 5-6.

5.10   STUB MATCHING

Either because a load has a reactive component or because a resistive load is not matched to the characteristic impedance of the line, the impedance seen from the input end of a transmission line can have a reactive component. In order to eliminate the standing waves on such a transmission line, it is necessary to eliminate the reactive component of impedance as seen looking into the line from the input end. One way to do this is to place a short section of transmission line, either open-circuited or short-circuited, in parallel with the transmission line at a location close to the load end of the line. Such a short section of

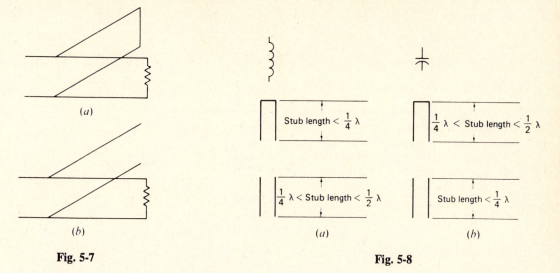

Fig. 5-7   Fig. 5-8

line is called a stub (see Fig. 5-7). Depending on the length of the short section of line and whether it is short-circuit-terminated or open-circuit-terminated, it will appear as either a capacitive reactance or an inductance reactance (see Fig. 5-8). The proper choice of stub length and its installation at an appropriate position on the transmission line will cause the input end of the line to be looking into a resistive condition, and at the design frequency (or harmonic of the design frequency) no standing waves will exist from the input end of the line to the place on the line where the stub is located. If the frequency of the signal changes, then the matched condition provided by the stub will no longer prevail and an unmatched condition with standing waves will again exist.

Generally stubs are less than one half-wavelength long. A short-circuit-terminated stub less than one quarter-wavelength long appears as an inductance, as does an open-circuit-terminated stub more than one quarter-wavelength long but less than one half-wavelength long. An open-circuit-terminated stub less than one quarter-wavelength long appears as a capacitance, as does a short-circuit-terminated stub whose length is more than one quarter-wavelength but less than one half-wavelength (see Fig. 5-8).

Although the determination of the length and location of the stub usually requires the use of the Smith chart and/or complex algebraic solutions, a simplified approach is possible using available graphs such as the one shown in Fig. 5-9.

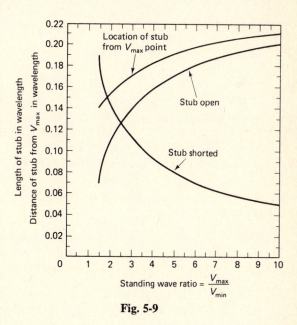

Fig. 5-9

Solved Problems

5.1 Determine the required pulse duration of a pulse so that when the pulse travels on a 25-meter line, the trailing edge occurs at the generator end of the line just as the leading edge reaches the load. Assume that the speed of the pulse on the line is the same as its free-space velocity (3×10^8 meters/s). See Fig. 5-10.

SOLUTION

Given: $d = 25$ meters

          $c = 3 \times 10^8$ meters/s

Find: t_p

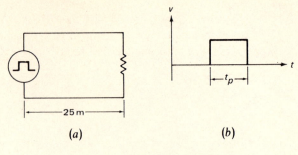

(a)   (b)

Fig. 5-10

The pulse duration for this situation should be equal to the amount of time it takes for the leading edge of the pulse to travel the 25-meter length of the line. Using the classic equation for rate, time, and distance,

$$\text{rate} \times \text{time} = \text{distance}$$
$$rt_p = d$$
$$(3 \times 10^8)t_p = 25$$
$$t_p = 8.33 \times 10^{-8} \text{ s}$$

$$\boxed{t_p = 0.0833 \ \mu s}$$

5.2   What is the wavelength of a 150-MHz sine wave? Make the determination first (a) in meters and then (b) in miles. Assume that the velocity of the wave on the line is the free-space velocity.

SOLUTION

Given: $f = 150$ MHz

Find: (a) λ in meters   (b) λ in miles

(a)   Assuming that the velocity of the wave on the transmission line is 3×10^8 meters/s, the wavelength in meters can be determined using

$$f\lambda = c$$
$$(150 \times 10^6)\lambda = 3 \times 10^8 \text{ meters/s}$$

Solving for λ,

$$\lambda = \frac{3 \times 10^8}{150 \times 10^6}$$

$$\boxed{\lambda = 2 \text{ meters}}$$

(b)   In order to determine the wavelength in miles, either the above answer can be converted from meters to miles, or the formula $f\lambda = c$ can be used again, this time taking c to be 186 000 miles/s.

$$f\lambda = c$$
$$150 \times 10^6 \lambda = 186\,000 \text{ miles/s}$$

Solving for λ,

$$\lambda = \frac{186\,000}{150 \times 10^6}$$

$$\boxed{\lambda = 0.001\,24 \text{ miles}}$$

5.3 A pulse train is transmitted along a transmission line which is 200 meters long. The pulse train consists of pulses with a duration of 30 ns each and separated by 45 ns. How many pulses can be on the line at any given time? Assume the speed of E/M waves to be the same as in free space. See Fig. 5-11.

SOLUTION

Given: $d = 200$ meters

$t_p = 30$ ns

$t_s = 45$ ns

Find: n

Fig. 5-11

$$t_T = t_p + t_s$$
$$= (30 \times 10^{-9}) + (45 \times 10^{-9})$$
$$= 75 \times 10^{-9}$$

The time required for one pulse and one space is 75 ns.
Now determine the space available on the line for a total combination of one pulse and one space.
Calculate the amount of time the pulse train remains on the line using

$$r \times t = d$$

Substitute numerical values and use 3×10^8 meters/s since the length of the line is given in meters.

$$3 \times 10^8 t = 200$$

Solving for t,

$$t = \frac{200}{3 \times 10^8}$$
$$= 66.667 \times 10^{-8}$$
$$= 666.67 \times 10^{-9}$$
$$t_L = 666.67 \text{ ns}$$

Thus the length of the line represents 666.67 ns.
Now find the number of pulse and separation combinations that fit on the line:

$$t_L = 666.67 \text{ ns}$$
$$t_T = t_d + t_s$$
$$= 30 \text{ ns} + 45 \text{ ns}$$
$$= 75 \text{ ns}$$
$$n_T = \frac{t_L}{t_T} = \frac{666.67 \times 10^{-9}}{75 \times 10^{-9}}$$
$$= 8.8889$$

Thus eight full pulses and space combinations fit on the line with some room left over. The next question is whether the remaining space is enough for another full pulse.
The eight combinations occupy

$$8 \times 75 \text{ ns} \qquad \text{or} \qquad 600 \text{ ns}$$

with

$$666.67 - 600 = 66.67 \text{ ns}$$

left over.
Since each pulse duration is 30 ns, there is room for 1 more whole pulse in addition to the 8 pulse and space combinations.
Thus there are a maximum of $8 + 1 = 9$ pulses on the line. So,

$$\boxed{n = 9}$$

5.4   How many 600-kHz waves can be on a 5-mile transmission line simultaneously?

SOLUTION

Given: $f = 600$ kHz

$\qquad\qquad d = 5$ miles

Find: n

First determine the wavelength of the 600-kHz signal using

$$f\lambda = c$$

Use 186 000 miles per hour as the speed of light since the line length is given in miles and no figure is given for the velocity factor.

Substituting numerical values,

$$600 \times 10^3 \lambda = 186\,000$$

$$\lambda = \frac{186\,000}{600 \times 10^3}$$

$$= 0.31 \text{ miles}$$

Knowing the wavelength of the signal and the length of the line, the number of cycles on the line can be found from

$$n = \frac{d}{\lambda}$$

$$= \frac{5}{0.31}$$

$$\boxed{n = 16.13}$$

5.5   A sine wave having a frequency of 75 MHz is launched on a transmission line.

(*a*) How long does it take from the time that the instantaneous voltage is zero before a peak occurs at the launch point?

(*b*) How far along the transmission line has the leading edge of the wavefront progressed in this amount of time?

Assume the speed of electromagnetic waves on this line to be the same as in free space (3×10^8 meters/s or 186 000 miles/s).

SOLUTION

Given: $f = 75$ MHz

Find: (*a*) $T/4$ (*b*) d

(*a*) It will take one quarter period for the wave to go from zero to a peak. See Fig 5-12. First determine the period and then determine one quarter of the period.

Determine the period T using the classical equation

$$T = \frac{1}{f}$$

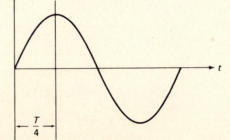

Fig. 5-12

Substituting numerical values,

$$T = \frac{1}{75 \times 10^6}$$

$$= 13.3 \times 15^{-9} \text{ s}$$

Thus the period is 13.3 ns.

One quarter of the full wavelength will provide the distance sought:

$$\frac{T}{4} = \frac{13.3 \times 10^{-9}}{4}$$

$$= 3.33 \times 10^{-9}$$

or

$$\boxed{\frac{T}{4} = 3.33 \text{ ns}}$$

(b) The distance that the leading edge has traveled in this amount of time can be calculated using the classical equation

$$d = r \times t$$

Substituting numerically,

$$d = (3 \times 10^8)(3.33 \times 10^{-9})$$

$$\boxed{d = 1.00 \text{ meter}}$$

5.6 Determine the characteristic impedance of a transmission line which has a capacitance of 35 pF/ft and an inductance of 0.25 μH/ft.

SOLUTION

Given: $c = 35$ pF/ft

$L = 0.25 \mu$H/ft

Find: Z_0

Using the equation relating Z_0, L, and C,

$$Z_0 = \sqrt{L/C}$$

Substituting numerical values,

$$Z_0 = \sqrt{\frac{0.25 \times 10^{-6}}{35 \times 10^{-12}}}$$

$$\boxed{Z_0 = 84.5 \ \Omega}$$

5.7 A particular cable has a capacitance of 40 pF/ft and a characteristic impedance of 70 Ω.

(a) What is the inductance per foot of this cable?

(b) Determine the impedance of an infinitely long section of such cable.

SOLUTION

Given: $C = 40$ pF/ft

$Z_0 = 70 \ \Omega$

Find: (a) L (b) Z

(a)   Using the formula relating characteristic impedance, capacitance, and inductance,

$$Z_0 = \sqrt{L/C}$$

Substituting numerical values,

$$70 = \sqrt{\frac{L}{40 \times 10^{-12}}}$$

Now, in order to get L out from under the radical sign, square both sides of the equation:

$$70^2 = \left[\sqrt{\frac{L}{40 \times 10^{-12}}} \right]^2$$

which results in

$$70^2 = \frac{L}{40 \times 10^{-12}}$$

Multiplying both sides of the equation by (40×10^{-12}), L can be obtained by itself on one side of the equals sign:

$$(70)^2 (40 \times 10^{-12}) = L$$

Rearranging,

$$L = 1.96 \times 10^{-7}$$
$$= 0.196 \times 10^{-6}$$

$$\boxed{L = 0.196 \, \mu H}$$

(b)   The characteristic impedance of a transmission line is the impedance that an infinite length of the line would present to a power supply at the input end of the line. So, in this case,

$$\boxed{Z_\infty = Z_0 = 70 \, \Omega}$$

5.8   Voltage and current readings are taken on a transmission line at different points. The maximum voltage reading is 60 $V_{rms\,max}$, and the minimum voltage reading is 20 $V_{rms\,min}$.

(a)   Calculate the VSWR on this line.

(b)   What is the ISWR on this line?

(c)   If the maximum current reading on the line is 2.5 A, what would the lowest current reading be?

SOLUTION

Given: $V_{rms\,max} = 60$ V
$V_{rms\,min} = 20$ V
$I_{rms\,max} = 2.5$ A

Find: (a) VSWR   (b) ISWR   (c) $I_{rms\,min}$

(a)   By definition,

$$VSWR = V_{rms\,max} : V_{rms\,min}$$
$$= 60 : 20$$

$$\boxed{VSWR = 3 : 1}$$

(b) Since   ISWR = VSWR,

$$\boxed{\text{ISWR} = 3\!:\!1}$$

(c)

$$\text{ISWR} = I_{rms\,max} : I_{rms\,min}$$

$$3\!:\!1 = 2.5 : I_{rms\,min}$$

Converting to fractions,

$$\frac{3}{1} = \frac{2.5}{I_{rms\,min}}$$

$$I_{rms\,min} = \frac{2.5}{3}$$

Thus,

$$\boxed{I_{rms\,min} = 0.833\ \text{A}}$$

5.9 A transmission line having a characteristic impedance of 75 Ω is delivering power to a 150-Ω load.

(a) Calculate the SWR on this line.

(b) Determine the minimum voltage reading on this line if the maximum voltage is 25 V.

SOLUTION

Given: $Z_0 = 75\ \Omega$

$Z_L = 150\ \Omega$

$V_{rms\,max} = 25\ \text{V}$

Find: (a) SWR   (b) $V_{rms\,min}$

(a) Since SWR, VSWR, and ISWR are all equal, determination of one provides all three.
From the equation relating SWR, Z_L, and Z_0,

$$\text{SWR} = \frac{Z_L}{Z_0}$$

$$= \frac{150}{75}$$

$$\boxed{\text{SWR} = 2\!:\!1}$$

(b)

$$\text{VSWR} = \text{SWR} = \frac{V_{rms\,max}}{V_{rms\,min}}$$

$$2 = \frac{25}{V_{rms\,min}}$$

$$V_{rms\,min} = \frac{25}{2}$$

$$\boxed{V_{rms\,min} = 12.5\ \text{V}}$$

5.10   A 50-Ω load is being fed from a 72-Ω transmission line. See Fig. 5-13.

   (a)   What is the standing-wave ratio resulting from this mismatch?

   (b)   Determine the reflection coefficient resulting from this mismatch.

   (c)   What percentage of the incident power is reflected from the load?

   (d)   What percentage of the incident power is absorbed by the load?

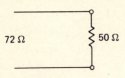

Fig. 5-13

SOLUTION

Given: $Z_0 = 72\ \Omega$

           $Z_L = 50\ \Omega$

Find: (a)   SWR   (b) K_r (c) $\dfrac{P_{\text{refl}}}{P_{\text{inc}}} \times 100$ (d) $\dfrac{P_{\text{abs}}}{P_{\text{inc}}} \times 100$

(a)   Setting up a ratio with the larger quantity taken first, the SWR can be found from

$$\text{SWR} = Z_0 : Z_L$$
$$= 72 : 50$$
$$= \frac{72}{50} : 1$$

$$\boxed{\text{SWR} = 1.44 : 1}$$

(b)   The reflection coefficient is related to Z_L and Z_0 by

$$K_r = \left| \frac{Z_L - Z_0}{Z_0 + Z_L} \right|$$
$$= \frac{72 - 50}{72 + 50}$$
$$= \frac{22}{122}$$

$$\boxed{K_r = 0.180}$$

(c)   Since K_r is a ratio of voltages at the load, K_r^2 is a ratio of powers:

$$\frac{P_{\text{refl}}}{P_{\text{inc}}} = K_r^2$$
$$= (0.180)^2 = 0.0324$$

Converting from a decimal to a percentage by multiplying by 100,

$$\% P_{\text{refl}} = 0.0324 \times 100$$

$$\boxed{\% P_{\text{refl}} = 3.24\%}$$

(d)   Knowing the percent of power reflected, the power absorbed is that left over from 100%.

$$\% P_{\text{abs}} = 100 - 3.24$$

$$\boxed{\% P_{\text{abs}} = 96.8\%}$$

5.11 (a) Determine the required length of a quarter-wave matching section which will eliminate standing waves and thereby provide a matched condition for a 300-Ω resistive load fed from a 72-Ω transmission line. This condition is to exist for a frequency of 100 MHz.

(b) Determine the characteristic impedance of the transmission line from which the matching section should be cut. Assume a velocity factor of 1.0.
See Fig. 5-14.

SOLUTION

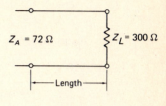

Given: $Z_L = 300\ \Omega$

$Z_A = 72\ \Omega$

$f = 100\ MHz$

Find: (a) Length

(b) Z_0

Fig. 5-14

(a) In order to determine the length of a quarter-wave matching section, it is first necessary to determine the wavelength of the 100-MHz signal:

$$f\lambda = C$$

$$\lambda = \frac{C}{f}$$

$$= \frac{3 \times 10^8}{100 \times 10^6}$$

$$= 3\ meters$$

$$Length = 0.25\,\lambda$$

$$= 0.25\,(3)$$

$$\boxed{Length = 0.75\ meter}$$

(b) The characteristic impedance of the required matching section is related to the load impedance and the original line by

$$Z_0 = \sqrt{Z_A Z_L}$$

Solving,

$$Z_0 = \sqrt{72(300)}$$

$$= \sqrt{21\,600}$$

$$\boxed{Z_0 = 146.97\ \Omega}$$

5.12 A 105-MHz, 90-V peak signal is incident on a 50-Ω transmission line. The velocity factor k of this line is 0.85. The line is 125 meters long and is terminated in a 300-Ω load. See Fig. 5-15.

(a) Find the wavelength of the signal on the line.

(b) Determine the length of the line in wavelengths.

(c) What is the SWR for this situation?

(d) Find the reflection coefficient.

(e) Calculate the peak value of the reflected voltage wave.

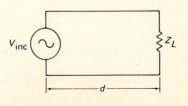

Fig. 5-15

(f)   What percent of the incident power is returned as reflected power?

(g)   Find the peak values of I_{inc} and I_{refl}.

(h)   Determine the peak value of the voltage standing wave at the voltage antinodes.

(i)   Determine the peak value of the current standing wave at the current antinodes.

(j)   Determine the peak value of the voltage standing wave at the voltage nodes.

(k)   Determine the peak value of the current standing wave at the current nodes.

(l)   If a quarter-wavelength matching section is to be used to correct for a mismatch, what must be its characteristic impedance?

SOLUTION

Given: $f = 105 \text{ MHz}$

$V_{inc\,peak} = 90 \text{ V}$

$Z_0 = 50 \, \Omega$

$k = 0.85$

$d = 125 \text{ meters}$

$Z_L = 300 \, \Omega$

Find: (a)   Wavelength   (e) $V_{refl\,peak}$ (i) $I_{peak\,total}$ at antinodes

(b) d in wavelengths   (f)   Percent power reflected   (j) $V_{peak\,total}$ at nodes

(c)   SWR   (g) $I_{inc\,peak}$, $I_{refl\,peak}$ (k) $I_{peak\,total}$ at nodes

(d) K_r (h) $V_{peak\,total}$ at antinodes   (l) Z_0

(a)   Taking into consideration the velocity factor k, the wavelength λ can be found from

$$f\lambda = kC$$

Substituting,

$$105 \times 10^6 \, \lambda = 0.85 \times 3 \times 10^8$$

$$\lambda = \frac{0.85 \times 3 \times 10^8}{105 \times 10^6}$$

$$\boxed{\lambda = 2.43 \text{ meters}}$$

(b)   Dividing the length of the line by the wavelength of the signal provides us with the length of the line in wavelengths:

$$\text{Wavelengths} = \frac{d}{\lambda} = \frac{125}{2.43}$$

Thus,

$$\boxed{d = 51.44 \text{ wavelengths}}$$

(c)   The standing-wave ratio is determined from Z_L and Z_0 by

$$\text{SWR} = Z_L : Z_0 = \frac{Z_L}{Z_0} : 1$$

Substituting,

$$\text{SWR} = \frac{300}{50} : 1$$

$$\boxed{\text{SWR} = 6 : 1}$$

(d)   The reflection coefficient can be found from

$$K_r = \left| \frac{Z_L - Z_0}{Z_L + Z_0} \right|$$

Substituting,

$$K_r = \frac{250}{350}$$

$$\boxed{K_r = 0.715}$$

(e)   Knowing the reflection coefficient and incident voltage, we can determine the peak value of the reflected voltage:

$$K_r = \frac{V_{\text{refl}}}{V_{\text{inc}}}$$

$$0.715 = \frac{V_{\text{refl peak}}}{90}$$

$$\boxed{V_{\text{refl peak}} = 64.35 \text{ V}}$$

(f)   Power relationships are equal to the square of voltage relationships for the same load $(P = V^2/R)$. Thus,

$$\frac{P_{\text{refl}}}{P_{\text{inc}}} = K_r^2 = (0.715)^2$$

So,

$$\% \text{ power reflected} = \frac{P_{\text{refl}}}{P_{\text{inc}}} \times 100 = K_r^2 \times 100$$

$$= (0.715)^2 \times 100$$

$$\boxed{\% \text{ power reflected} = 51\%}$$

(g)   Applying Ohm's law at any point on the transmission line,

$$\frac{V_{\text{inc}}}{I_{\text{inc}}} = Z_0$$

$$I_{\text{inc peak}} = \frac{90}{50}$$

$$\boxed{I_{\text{inc peak}} = 1.8 \text{ A}}$$

Similarly,

$$\frac{V_{\text{refl}}}{I_{\text{refl}}} = Z_0$$

$$I_{\text{refl peak}} = \frac{64.35}{50}$$

$$\boxed{I_{\text{refl peak}} = 1.287 \text{ A}}$$

(h) The antinodes are those points on the transmission line at which the incident and reflected waves totally reinforce each other:

$$V_{\text{peak total max}} = V_{\text{inc peak}} + V_{\text{refl peak}}$$
$$= 90 + 64.35$$

$$\boxed{V_{\text{peak total max}} = 154.35 \text{ V}}$$

(i)

$$I_{\text{peak total max}} = I_{\text{inc peak}} + I_{\text{refl peak}}$$
$$= 1.8 + 1.287$$

$$\boxed{I_{\text{peak total max}} = 3.087 \text{ A}}$$

(j) The nodes are those points on the transmission line at which the incident and reflected waves are in maximum opposition to each other.

$$V_{\text{peak total min}} = V_{\text{inc peak}} - V_{\text{refl peak}}$$
$$= 90 - 64.35$$

$$\boxed{V_{\text{peak total min}} = 25.65 \text{ V}}$$

(k)

$$I_{\text{peak total min}} = I_{\text{inc peak}} - I_{\text{refl peak}}$$
$$= 1.8 - 1.287$$

$$\boxed{I_{\text{peak total min}} = 0.513 \text{ A}}$$

(l) The characteristic impedance of the matching section can be found from

$$Z_0 = \sqrt{Z_A Z_L}$$

Substituting,

$$Z_0 = \sqrt{50(300)}$$
$$= \sqrt{15\,000}$$

$$\boxed{Z_0 = 122.47 \ \Omega}$$

5.13 A short-circuited stub is required for use in a situation in which there is an SWR of 4 : 1 on a transmission line. Determine the length of the stub and its appropriate location on the line. The frequency at which standing waves are to be eliminated is 200 MHz.

SOLUTION

Given: SWR = 4:1
$$f = 200 \text{ MHz}$$

Find: L_s, d_s

Using Fig. 5-9 and entering the graph at SWR = 4, it is seen that the length of the shorted stub must equal 0.09λ and be placed 0.181λ from the voltage antinode. The wavelength of the 100-MHz signal is

$$f\lambda = 3 \times 10^8$$
$$200 \times 10^6 \, \lambda = 3 \times 10^8$$
$$\lambda = \frac{3 \times 10^8}{200 \times 10^6}$$
$$\lambda = 1.5 \text{ meters}$$

Stub length is thus

$$L_s = 0.09(1.5)$$

$$\boxed{L_s = 0.135 \text{ meter}}$$

The distance from the voltage antinode is

$$d_s = 0.181(1.5)$$

$$\boxed{d_s = 0.2715 \text{ meter}}$$

Supplementary Problems

5.14 A pulse has a duration of 20 μs. How far down the transmission line does the leading edge get by the time the trailing edge appears at the generator end of the line?   *Ans.*   6000 meters

5.15 Determine the pulse duration of a pulse such that its trailing edge occurs at the generator just as its leading edge gets halfway down a 10-meter transmission line. Assume the speed of E/M waves on the line to be the same as in free space.   *Ans.*   16.67 ns

5.16 A transmission line is 80 meters long. The line is being fed a pulse train consisting of pulses each having a duration of 15 ns and spaced 25 ns apart. How many pulses can be on the line at any given time? Assume that the speed of E/M waves on the line is the same as in free space.   *Ans.*   7 pulses

5.17 A 3-mile transmission line is fed a pulse train consisting of pulses each having a duration of 40 ns and separated from each other by 70 ns. How many pulses can be on the line at any given time?   *Ans.*   147

5.18 How many 1-MHz waves can be on a 4-mile transmission line simultaneously?   *Ans.*   21.5

5.19 On a transmission line 4.5 cycles of a 2.5-MHz signal appear simultaneously. How long is the line in miles and in meters?   *Ans.*   0.3348 mile, 540 meters

5.20 A transmission line is 500 meters long. How many pulses of a pulse train are on the line simultaneously? The pulse duration is 150 ns while rest time between pulses is 250 ns.   *Ans.*   4+

5.21 Calculate the wavelength in free space of the following electromagnetic signals in meters, in miles, and in feet.

(*a*) 20 Hz	(*c*) 2000 Hz	(*e*) 200 000 Hz	(*g*) 20 MHz
(*b*) 200 Hz	(*d*) 20 000 Hz	(*f*) 2 MHz	(*h*) 200 MHz

Ans.   Meters: (*a*)   15 000 k   (*c*)   150 k   (*e*)   1.5 k   (*g*)   15
                 (*b*)   1500 k   (*d*)   15 k   (*f*)   150   (*h*)   1.5

           Miles: (*a*)   9300   (*c*)   93   (*e*)   0.93   (*g*)   0.0093
                 (*b*)   930   (*d*)   9.3   (*f*)   0.093   (*h*)   0.000 93

            Feet: (*a*)   49 104 000   (*c*)   491 040   (*e*)   4910   (*g*)   49.1
                 (*b*)   4 910 400   (*d*)   49 104   (*f*)   491   (*h*)   4.91

5.22 A cable has a capacitance of 10 pF/ft and an inductance of 0.03 μH/ft. What is the characteristic impedance of the cable?   *Ans.*   54.77 Ω

5.23 Determine the characteristic impedance of a transmission line which has a capacitance of 12 pF/ft and an inductance of 0.015 μH/ft.   *Ans.*   35.36 Ω

5.24   Calculate the characteristic impedance of a cable having a capacitance of 25 pF/ft and an inductance of
0.18 μH/ft.   *Ans.*   84.85 Ω

5.25   Find the characteristic impedance of a transmission line that has a capacitance of 20 pF/ft and an inductance of
0.22 μH/ft.   *Ans.*   104.88 Ω

5.26   Find the capacitance per foot of a transmission line having a characteristic impedance of 80 Ω and an inductance
of 0.30 μH/ft.   *Ans.*   46.875 pF

5.27   Calculate the capacitance per foot of a cable that has a characteristic impedance of 75 Ω and an inductance of
0.19 μH/ft.   *Ans.*   33.778 pF

5.28   A cable has a capacitance of 40 pF/ft and a characteristic impedance of 300 Ω. What is the inductance per foot of
this cable?   *Ans.*   3.6 μH

5.29   Find the inductance per foot of a transmission line that has a characteristic impedance of 200 Ω and a capacitance
of 35 pF/ft.   *Ans.*   1.4 μH

5.30   Determine the standing-wave ratio on a transmission line on which the maximum rms voltage is 90 V and the
minimum rms voltage is 25 V.   *Ans.*   3.6:1

5.31   What is the SWR on a transmission line which has a maximum rms voltage of 118 V and a minimum rms voltage
of 50 V?   *Ans.*   2.36:1

5.32   A 1.5-MHz sine wave is launched on a transmission line. How much time does it take from the instant that the
instantaneous voltage is zero until a peak occurs at the launch point? How far along the transmission line has the
leading edge of the wavefront progressed in this amount of time?

           Assume the speed of electromagnetic waves on this line to be the same as in free space (3×10^8 meters/s or
186 000 miles/s).   *Ans.*   166.67 ns, 50 meters

5.33   A 40-MHz sine wave is transmitted on a transmission line. How much time elapses between the occurrence of
instantaneous zero values at the launch point? How far down the transmission line has the first incident instantaneous
zero value progressed when the second zero occurs at the launch point?   *Ans.*   12.5 ns, 3.75 meters

5.34   Find the SWR on a transmission line having a maximum rms current of 1.75 A and a minimum of 0.85 A at different
points on the line.   *Ans.*   2.059:1

5.35   Voltage and current readings are taken at many different points on a transmission line. The maximum voltage reading
is 140 V rms and a minimum voltage reading of 65 V rms. The maximum current reading on the line is 4.8 A.

   (*a*)   Calculate the VSWR on this line.

   (*b*)   What is the ISWR on this line?

   (*c*)   Determine the lowest current reading on the line.

   Ans.   (*a*) 2.1538:1, (*b*) 2.1538:1, (*c*) 2.229 A

5.36   The maximum rms current reading anywhere along a particular transmission line is 8.4 A while the minimum rms
current reading anywhere on the line is 2.8 A.

   (*a*)   What is the ISWR on this line?

   (*b*)   Determine the VSWR on this line.

   (*c*)   If the maximum rms voltage on this line is 178 V, what is the minimum rms voltage on the line?

   Ans.   (*a*) 3:1, (*b*) 3:1, (*c*) 59.33 V

5.37   A 250-Ω transmission line carries a signal from a source to a 40-Ω load. Determine the standing-wave ratio on the line.   *Ans.*   6.25:1

5.38   A 50-Ω transmission line delivers energy to a 300-Ω source. What is the SWR on the line?   *Ans.*   6:1

5.39   A 300-Ω transmission line delivers energy to a 65-Ω load. Calculate the SWR on the line.   *Ans.*   4.6:1

5.40   A 72-Ω transmission line has a standing-wave ratio of 8:1 due to a mismatch between the line and the resistive load. What is the resistance of the load?   *Ans.*   9 Ω or 576 Ω

5.41   A 4.3:1 standing-wave ratio appears on a transmission line due to a mismatch between the line impedance and the impedance of a resistive load. Determine the resistance of the load if the characteristic impedance of the line is 140 Ω.   *Ans.*   32.56 Ω or 602 Ω

5.42   A standing-wave ratio of 2:1 results when a certain transmission line is connected to a 75-Ω resistive load. What is the characteristic impedance of the transmission line?   *Ans.*   150 Ω or 37.5 Ω

5.43   An SWR of 3:1 results when a transmission line is terminated in a 90-Ω load. Determine the characteristic impedance of the transmission line.   *Ans.*   30 Ω or 270 Ω

5.44   What is the characteristic impedance of a transmission line if terminating it in a 300-Ω load results in an SWR of 2:1?   *Ans.*   150 Ω or 600 Ω

5.45   Calculate the reflection coefficient on a transmission line having an SWR of 3:1.   *Ans.*   .5

5.46   Determine the reflection coefficient on a transmission line having an SWR of 2.5:1.   *Ans.*   .429

5.47   (*a*)   Determine the SWR and the reflection coefficient for a transmission line whose characteristic impedance is 50 Ω and which is terminated in a 90-Ω resistive load.

       (*b*)   What percentage of the incident power is reflected and what percentage is absorbed by the load?

    Ans.   (*a*) 1.8:1, .286; (*b*) 8.16%, 91.84%

5.48   (*a*)   What SWR and reflection coefficient result when a 300-Ω transmission line is used to feed a 70-Ω load?

       (*b*)   What percentage of the incident power is reflected by the load and what percentage is absorbed by the load?

    Ans.   (*a*) 4.286:1, .622; (*b*) 38.6%, 61.4%

5.49   What is the necessary length and characteristic impedance of a cable to be used as a quarter-wave matching transformer which is to eliminate the standing waves due to the mismatch between a 50-Ω transmission line and a 300-Ω load? The match is desired at 60 MHz. Assume a velocity factor of 1.0.
    Ans.   1.25 meters, 122.47 Ω

5.50   Calculate the length and characteristic impedance of a transmission line which is to act as a quarter-wave matching transformer between a 180 Ω transmission line and a 400-Ω load. The matching action is to be effective at 600 kHz. Assume a velocity factor of 1.0.   *Ans.*   125 meters, 268.33 Ω

5.51   Find the required length and characteristic impedance of a quarter-wave matching transformer to be used to eliminate standing waves due to a mismatch between a 135-Ω transmission line and a 40-Ω resistive load at 144 MHz. Assume a velocity factor of 1.0.   *Ans.*   0.521 meter, 73.48 Ω

5.52   A 50-Ω cable that is 0.7 meter long is being used to match a 300-Ω transmission line to a load. Determine the resistance of the load and the frequency at which the match is supposed to occur. Assume a velocity factor of 1.0.   *Ans.*   8.33 Ω, 107 MHz

5.53   Determine the appropriate length for a quarter-wave matching section to be effective at 55 MHz if the velocity factor
on this line is 0.85.   *Ans.*   1.159 meters

5.54   For what frequency is a 2.3-meter quarter-wave matching section cut if the velocity factor on this line is 0.92?
Ans.   30 MHz

5.55   A 250-MHz 40-V peak signal is incident on a 72-Ω transmission line. The velocity factor for this line is 0.91. The
line is 250 meters long and is terminated in a 200-Ω load.

(*a*)   Find the wavelength of the signal on the line.

(*b*)   Determine the length of the line in wavelengths.

(*c*)   What is the SWR for this situation?

(*d*)   Find the reflection coefficient.

(*e*)   Calculate the peak value of the reflected voltage wave.

(*f*)   What percent of the incident power is returned as reflected power?

(*g*)   Find the peak values of I_{inc} and I_{refl}.

(*h*)   Determine the peak value of the voltage standing wave at the voltage antinodes.

(*i*)   Determine the peak value of the current standing wave at the current antinodes.

(*j*)   Determine the peak value of the voltage standing wave at the voltage nodes.

(*k*)   Determine the peak value of the current standing wave at the current nodes.

(*l*)   If a quarter-wavelength matching section is to be used to correct for a mismatch, what must its characteristic
impedance be?

Ans.   (*a*) 1.092 meters; (*b*) 228.94 wavelengths; (*c*) 2.778:1; (*d*) 0.4706; (*e*) 18.82 V; (*f*) 22.146%; (*g*) 0.5556 A,
0.2614 A; (*h*) 58.82 V; (*i*) 0.817 A; (*j*) 21.18 V; (*k*) 0.2942 A; (*l*) 120 Ω

5.56   A 112.6-MHz 75-V peak signal is incident on a 150-Ω transmission line. The velocity factor for this line is 0.89.
This line is 175 meters long and is terminated in a 70-Ω load.

(*a*)   Find the wavelength of the signal on the line.

(*b*)   Determine the length of the line in wavelengths.

(*c*)   What is the SWR for this situation?

(*d*)   Find the reflection coefficient.

(*e*)   Calculate the peak value of the reflected voltage wave.

(*f*)   What percent of the incident power is returned as reflected power?

(*g*)   Find the peak values of I_{inc} and I_{refl}.

(*h*)   Determine the peak value of the voltage standing wave at the voltage antinodes.

(*i*)   Determine the peak value of the current standing wave at the current antinodes.

(*j*)   Determine the peak value of the voltage standing wave at the voltage nodes.

(*k*)   Determine the peak value of the current standing wave at the current nodes.

(*l*)   If a quarter-wavelength matching section is to be used to correct for a mismatch, what must its characteristic
impedance be?

Ans.   (*a*) 2.37 meters; (*b*) 73.84 wavelengths; (*c*) 2.143:1; (*d*) 3636; (*e*) 27.27 V; (*f*) 13.22%; (*g*) 0.5 A, 0.1818 A;
(*h*) 102.27 V; (*i*) 0.6818 A; (*j*) 47.73 V; (*k*) 0.3182 A; (*l*) 102.47 Ω

5.57   Determine the length of a short-circuited stub and its location in order to eliminate an SWR of 3:1 on a transmission
line handling a 75-MHz signal. Assume a velocity factor of 1.0.   *Ans.*   0.448 meter, 0.68 meter

5.58   What is the proper length and location of a short-circuited stub in order for it to eliminate a 2.5:1 SWR at a frequency
of 144 MHz? Assume a velocity factor of 0.88 for the line.   *Ans.*   0.238 meter, 0.2988 meter

Chapter 6

Antennas

INTRODUCTION

An antenna is a device whose function is to radiate electromagnetic energy and/or intercept electromagnetic radiation. A transmitting antenna can be used for reception and vice versa. In two-way communications, the same antenna is used for both transmission and reception.

6.1   RADIATION PATTERNS

Antennas do not necessarily perform equally well in all directions. A polar diagram which indicates how well an antenna transmits or receives in different directions is called a radiation pattern.

Figure 6-1 is a radiation pattern for an antenna known as a *Marconi antenna.*

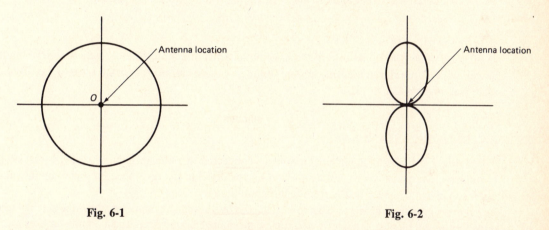

| Fig. 6-1 | Fig. 6-2 |

Figure 6-2 is a radiation pattern for an antenna known as a *Hertz antenna.* The distance from the location of the antenna to a point on the radiation pattern indicates the relative strength of the radiation in the direction determined by these two points.

Once a radiation-pattern diagram exists, as in Fig. 6-3, to determine the relative strength of the signal at point *A*, a line is drawn from the origin, point *O*, to point *A*. The intercept of this line with the radiation pattern determines the termination of the vector. When an antenna is being used for receiving purposes, the radiation pattern becomes a *reception pattern.* The long section of the lobes of the pattern indicates the best direction for reception.

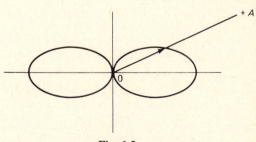

Fig. 6-3

6.2   BEAM WIDTH

It is frequently necessary to have a quick means of comparing the directivity of antennas without going through a point-by-point comparison of their radiation patterns. The beam width is such a quick description. The beam width of an antenna is the angle within which the power radiated is above one-half of what it is in the most preferential direction, or it can be said that the beam width is the angle within which the voltage developed by a receiving antenna remains within 70.7% of the voltage developed by

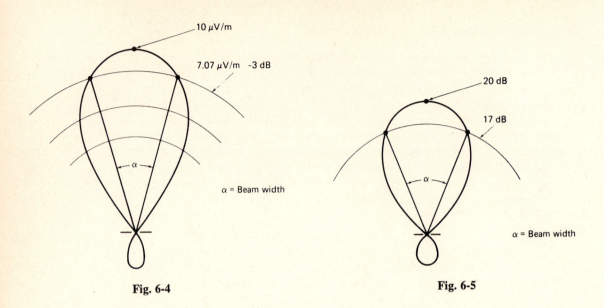

Fig. 6-4   Fig. 6-5

an antenna when it is aimed toward the most preferential direction. Another way of describing the half power points is to refer to them as the 3-dB points, because half power corresponds to -3 dB on the dB scale (see Figs. 6-4 and 6-5).

6.3   ANTENNA RESISTANCE

If power is to get to an antenna, it must be connected to a transmission line. It is necessary that the characteristic impedance of the transmission line be equal to the resistance presented by the antenna in order to prevent standing waves from being present on the line.

If measuring the antenna resistance is attempted by putting an ohmmeter across the antenna terminals, a reading would be obtained which would indicate an open circuit because an ohmmeter uses a dc source in measuring resistance. Therefore something other than that measured by an ohmmeter must be involved. The resistance presented by the antenna consists mainly of what is called *radiation resistance*.

The radiation resistance of an antenna is defined as a fictitious resistance which would dissipate as much power as the antenna in question is radiating if it were connected to the same transmission line (see Fig. 6-6). An antenna which is radiating 100 W when drawing 2 A has a radiation resistance of $100/2^2$, or 25 Ω ($P/I^2 = R$). Do not lose sight of the fact that this is not a true resistance. A true resistance causes heat losses. There *are* heat losses involved in an antenna. The heat losses are not what is accounted for by the radiation resistance. The radiation resistance accounts for the power which is radiated. There is another resistance associated with an antenna which is defined to take this heat loss into account. This is called the *ohmic resistance* of the antenna. It represents the actual losses caused by the conversion of electrical energy to heat as a result of the resistivity of the various conducting elements of the antenna.

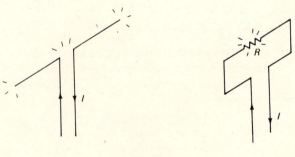

Fig. 6-6

6.4   THE ANTENNA AS A RESONANT CIRCUIT

The impedance presented by an antenna also has a reactive component due to currents and voltages being out of phase. The reason that currents and voltages are out of phase can be due to the antenna not being cut to the exact length prescribed for the type of antenna being considered. Actually an antenna is very much like a tuned circuit in that at its center frequency, that frequency at which its geometry is exactly correct, a maximum impedance which is purely resistive (radiation resistance plus ohmic resistance) is presented to the transmission line. As the frequency of the signal being presented to the antenna is changed, the impedance presented to the transmission line becomes reactive, just as with a tuned circuit. The bandwidth and Q of an antenna can be discussed just as is done with tuned circuits, and these terms still maintain their original meaning. The relationship between the bandwidth and the Q of an antenna is the same as that for tuned circuits.

$$\text{Bandwidth} = \frac{f_0}{Q} \tag{6.1}$$

If for any reason the resonant frequency of an antenna is not exactly at the frequency to be transmitted, the antenna can be tuned. *Antenna tuners* are used for this purpose. Antenna tuners are tunable reactive circuits which are tuned so that when combined with the antenna, the reactive components of the antenna and the tuning circuit cancel out. Capacitive reactance can counterbalance inductive reactance, and inductive reactance can counterbalance capacitive reactance.

6.5   VELOCITY FACTOR

As seen in Chapter 5, electromagnetic waves do not travel at the same speed in all media. The figures 186,000 miles/s or 3×10^8 meters/s for the speed of electromagnetic radiation are valid only for free space.

The velocity of electromagnetic waves is slightly different in a conductor such as the material the antenna is made of: aluminum, copper, etc. As mentioned before, in order to take this into account, the velocity factor is used. The velocity factor is that number which when multiplied by the speed of light in free space gives us the speed of light in the medium in question.

6.6   ANTENNA TYPES

The Half-Wave Dipole and the Marconi Antenna

Two of the more basic antennas are the half-wave dipole or Hertz antenna and the quarter-wave vertical or Marconi antenna. The half-wave dipole is shown as Fig. 6-7, while the quarter-wave vertical is shown as Fig. 6-8.

Fig. 6-7   Fig. 6-8

As is obvious from their names, the optimum length for each of the two antennas is one half wavelength for the Hertz and one quarter wavelength for the Marconi antenna.

Beam Antennas

A beam antenna is an antenna which has highly directional properties, essentially radiating a beam of electromagnetic radiation.

The *Yagi-Uda antenna* is frequently referred to as a beam antenna because of its highly directional radiation pattern.

There are many other antennas which have highly directional characteristics and are considered beam antennas. One of these is the *rhombic antenna*, which is shown in Fig. 6-9. It is constructed in a horizontal plane and has a radiation pattern as shown. The input resistance of the rhombic antenna is 800 Ω.

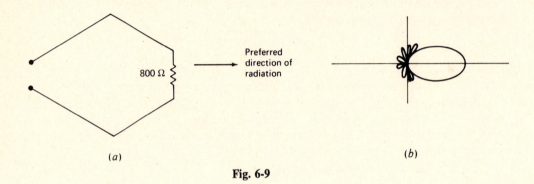

Fig. 6-9

Folded Dipole Antenna

While we are discussing other antennas, let us consider the folded dipole, shown in Fig. 6-10. The folded dipole is really a variation of the dipole antenna; the folded dipole has an input impedance of approximately 280 Ω and can be used quite conveniently with flat ribbon-type transmission line (twinex), which has a characteristic impedance of 300 Ω. The folded dipole with a reflector and twinex transmission line is quite popular for use with home-type television receivers.

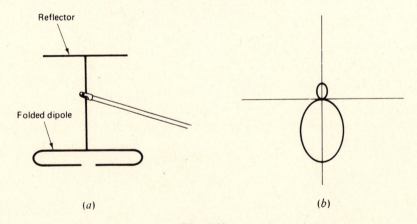

Fig. 6-10

Turnstile Antenna

An interesting variation on the dipole antenna is called the turnstile antenna. Recall the figure-eight, double-lobe pattern of the dipole antenna. Consider the radiation pattern that would result if two dipoles were constructed on the same mast perpendicular to each other, one radiating preferentially in the north-south direction and the other radiating preferentially in the east-west direction. A double figure-eight would present itself, and where the two overlapped we would have to do vector addition of the two fields; the result would be an almost circular radiation in the horizontal plane (see Fig. 6-11). The radiation patterns indicated are based upon the two dipoles being fed by signals 90° out of phase with each other but identical in all other respects. The turnstile antenna has an input resistance of 36 Ω (1/2 of that of a dipole, 72 Ω).

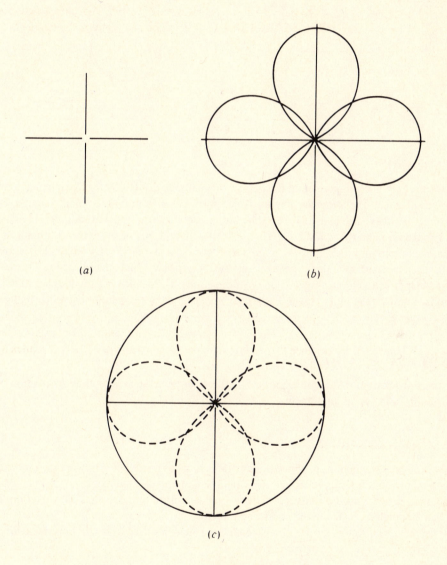

(a)

(b)

(c)

Fig. 6-11

Loop Antenna

An antenna which is used very frequently but almost entirely as a reception antenna is the loop antenna. This is the antenna which is frequently found on the back of ac-dc table radios. It can be a square loop or an oblong loop. The number of turns of wire can be anywhere from one to dozens in a coil pasted to the inside of the cabinet or to the fiberboard back. This antenna and its radiation pattern are shown in Fig. 6-12.

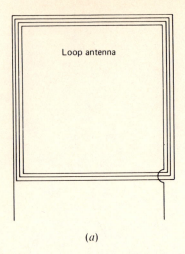

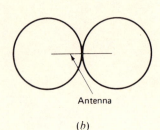

(a)   (b)

Fig. 6-12

6.7   ANTENNA GAIN

Antenna gain is a comparison of the output, in a particular direction, of the antenna in question and a *reference antenna*. The reference antenna is generally either an omnidirectional antenna, of which the Marconi quarter-wave antenna is an example, putting out equal amounts of radiation in all directions (a circular radiation pattern), or a dipole. Therefore, if it is said that an antenna has a gain of 10 dB, the antenna in question improves upon the reference antenna in that direction by 10 dB. The increased power being radiated in a particular direction is obtained at the expense of the other directions. Power is radiated in a particular direction by stealing it from other directions. Thus, antenna gain does *not* refer to obtaining more output power than input power. Frequently one antenna is compared to another, thus avoiding the need for a reference.

6.8   FRONT-TO-BACK RATIO

Front-to-back ratio is the ratio expressed in dB of output in the most optimum direction to the output 180° away from the optimum direction.

6.9   REFLECTORS AND DIRECTORS

There are many situations in which it is desirable to focus the radiated power into a more limited area than is possible with the simple dipole. For example, in radio communications between two stations, it is desirable to concentrate the total radiated power of the transmitting station in one direction. This desired effect can be obtained by using reflectors and directors.

Reflectors and directors are additional conducting elements used to obtain improved directivity of an antenna. The director is placed in front of the driven element (the dipole), while the reflector is placed behind the driven element.

Figure 6-13 is a drawing of an antenna which consists of one dipole, one reflector, and one director. Note that the director is less than one half-wavelength long and is placed closer to the dipole than is the reflector. The optimum spacing between the dipole and the director is 0.1 wavelength when the length of the director is 5% smaller than that of the dipole.

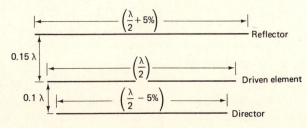

Fig. 6-13

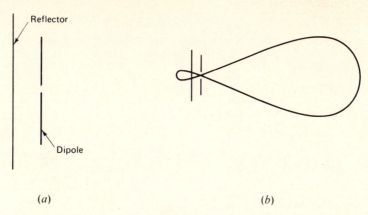

(a)   (b)

Fig. 6-14

The reflector is slightly longer than one half-wavelength and is placed less than one quarter-wavelength behind and parallel to the dipole. The optimum spacing is 0.15 wavelength if the reflector is 5% longer than the dipole. Figure 6-14 shows the radiation pattern of the dipole and reflector combination. Note the increased radiation in one direction at the expense of the other directions.

Directors and reflectors are known as *parasitic elements* because when used in a transmitting antenna they are not connected to the transmission line, but get their energy from the power transmitted by the dipole, which is connected to the transmission line. The dipole is known as the *driven element*. An antenna which consists of a driven element and any number of parasitic elements is known as a *parasitic array*. A dipole with reflectors and directors is frequently called a Yagi-Uda antenna, or simply a Yagi antenna.

6.10 ANTENNA TRAPS

There are situations in which one station may transmit at different frequencies at different times, making it desirable to be able to radiate efficiently at each of these frequencies. One possibility is to make use of the same transmission line and antenna but lengthen the antenna when signals of longer wavelength are to be transmitted. This can be done by placing LC circuits at various points on the antenna. These LC circuits have an impedance which ranges from very low to very high values. By choosing the L's and C's and the Q of the combination properly, a unit called a trap can be designed which behaves as an open circuit over a range of frequencies, thereby removing the portion of the antenna between it and the end, the only section of the antenna which is active being that portion between the driving point and the traps. In a center-fed dipole, traps are encountered in pairs, one in each of the quarter-wavelength sections, and the antenna length is taken as the distance between the two traps when the antenna is operating at frequencies at which the traps are considered open circuits (see Fig. 6-15).

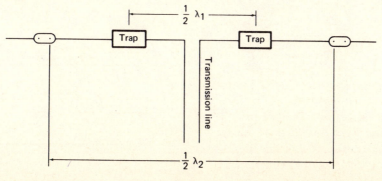

Fig. 6-15

Traps are commercially available for different frequencies. It is acceptable to have more than one set of traps on a line so that the antenna length can change to take on many different values.

6.11　WAVE PROPAGATION

Once a radiated signal leaves the antenna, it travels along one of three routes:

1. Along the ground (the *ground wave*)

2. Straight out in a straight line (the *line-of-sight wave*)

3. Up to the ionosphere and back to earth (the *sky wave*)

See Fig. 6-16.

The frequency of the signal determines which of these modes predominates. See Fig. 6-17.

The effectiveness of the line-of-sight wave is, as its name implies, limited to a line of sight between the transmitting and receiving antennas.

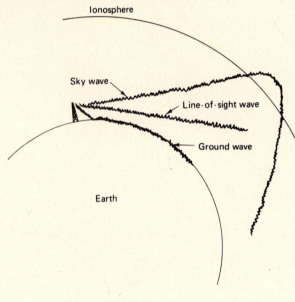

Fig. 6-16

An equation which can be used to calculate the maximum distance between transmitting and receiving antennas for direct line-of-sight waves to be effective is

$$d = \sqrt{2h_t} + \sqrt{2h_r}$$

where h_t is the height of the transmitting antenna in feet, h_r is the height of the receiving antenna in feet, and d is the maximum distance in miles over which communication between them can take place by direct line-of-sight wave.

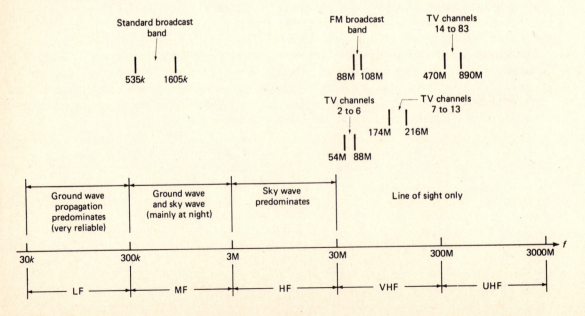

Fig. 6-17

Solved Problems

6.1   How much power will an antenna having a radiation resistance of 50 Ω radiate when it is fed 20 A?

SOLUTION

Given: $I = 20$ A

               $R_{rad} = 50$ Ω

Find: P_{rad}

   Using the basic power equation,

$$P = I^2 R$$
$$= (20)^2 (50)$$
$$\boxed{P = 20{,}000\,\text{W}}$$

6.2   What is the radiation resistance of an antenna which radiates 5 kW when it draws 15 A?

SOLUTION

Given: $P = 5$ kW

               $I = 15$ A

Find: R_{rad}

   From the basic power equation,

$$P = I^2 R$$
$$5000 = (15)^2 R_{rad}$$
$$R_{rad} = \frac{5000}{(15)^2}$$
$$\boxed{R_{rad} = 22.2\,\Omega}$$

6.3   An antenna having a radiation resistance of 75 Ω is radiating 10 kW. How much current flows into the antenna?

SOLUTION

Given: $R_{rad} = 75$ Ω

               $P_{rad} = 10$ kW

Find: I

   Using the basic power equation,

$$P = I^2 R$$
$$10 \times 10^3 = I^2 (75)$$
$$I^2 = \frac{10 \times 10^3}{75}$$
$$= 133.33$$
$$\boxed{I = 11.547\,\text{A}}$$

6.4   Determine the Q of an antenna if it has a bandwidth of 0.6 MHz and is cut to a frequency of 30 MHz.

SOLUTION

Given: BW = 0.6 MHz

f_0 = 30 MHz

Find: Q

Using the BW equation,

$$\text{BW} = \frac{f_0}{Q}$$

$$0.6 \times 10^6 = \frac{30 \times 10^6}{Q}$$

$$Q = \frac{30 \times 10^6}{0.6 \times 10^6}$$

$$\boxed{Q = 50}$$

6.5   Determine the length of a half-wave dipole antenna to be used to receive a 5-MHz radio signal. Assume that the velocity of electromagnetic waves on the antenna is 3×10^8 meters/s. See Fig. 6-18.

SOLUTION

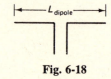

Fig. 6-18

Given: $f = 5$ MHz

Dipole antenna

$C = 3 \times 10^8$ m/s

Find: L_{dipole}

The length of a dipole antenna is equal to one half-wavelength. Finding the wavelength of the signal on the antenna is the first order of business:

$$f\lambda = C$$

$$5 \times 10^6 \lambda = 3 \times 10^8$$

$$\lambda = \frac{3 \times 10^8}{5 \times 10^6}$$

$$= 60 \text{ m}$$

The wavelength must now be divided by 2 in order to determine the required length of the dipole.

$$L_{\text{dipole}} = \frac{\lambda}{2} = \frac{60}{2}$$

$$\boxed{L_{\text{dipole}} = 30 \text{ m}}$$

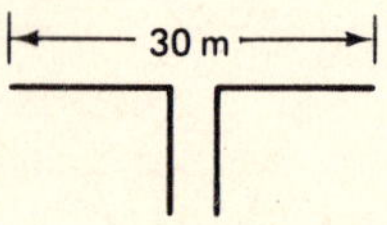

See Fig. 6-19.

|← 30 m →|

Fig. 6-19

6.6   A half-wave dipole antenna is to be cut to optimally radiate a 250-MHz signal. The velocity factor of the antenna elements is 0.85. Determine the necessary length of the antenna when taking the velocity factor into account, and then determine what the length would be if the velocity factor were 1.0.

SOLUTION

Given: $f = 250\,\text{MHz}$

$k_1 = 0.85$

$k_2 = 1.0$

Find: $L_1, \quad L_2$

$$f\lambda_1 = k_1 C$$
$$250 \times 10^6\,\lambda_1 = 0.85(3 \times 10^8)$$
$$\lambda_1 = \frac{0.85(3 \times 10^8)}{250 \times 10^6}$$
$$= 1.02\,\text{m}$$
$$L_1 = \frac{\lambda_1}{2}$$
$$= \frac{1.02}{2}$$

$$\boxed{L_1 = 0.51\,\text{m}}$$

$$f\lambda_2 = k_2 C$$
$$250 \times 10^6\,\lambda_2 = (1)(3 \times 10^8)$$
$$\lambda_2 = \frac{3 \times 10^8}{250 \times 10^6}$$
$$= 1.20\,\text{m}$$
$$L_2 = \frac{\lambda_2}{2}$$
$$= \frac{1.2}{2}$$

$$\boxed{L_2 = 0.6\,\text{m}}$$

6.7   An antenna is needed for the transmission of a signal having a center frequency of 60 MHz. Determine the length of a Hertz dipole antenna suited for this purpose. Assume a velocity factor of 0.85 for the antenna conductors. See Fig. 6-20.

SOLUTION

Given: $f = 60\,\text{MHz}$

$k = 0.85$

Find: L_{dipole}

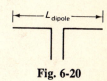

Fig. 6-20

The length of a Hertz dipole antenna should be equal to half the wavelength of the signal on the antenna. Thus we must find the wavelength of the signal on the antenna.

$$f\lambda = kC$$
$$60 \times 10^6\,\lambda = 0.85(3 \times 10^8)$$
$$\lambda = \frac{0.85(3 \times 10^8)}{60 \times 10^6}$$
$$= 0.0425 \times 10^2$$
$$= 4.25\,\text{m}$$

The length of the Hertz dipole is $\lambda/2$. Thus,

$$\frac{\lambda}{2} = \frac{4.25}{2}$$

$$\boxed{\frac{\lambda}{2} = 2.125\text{ m}}$$

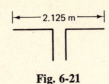

Fig. 6-21

See Fig. 6-21.

6.8 Determine the optimum length of a Marconi antenna for the transmission of a 100-MHz signal. The velocity factor for the antenna is 0.90. See Fig. 6-22.

SOLUTION

Given: $f = 100$ MHz

           $k = 0.90$

Find: L_{Marconi}

The optimum length of a Marconi vertical antenna is one quarter-wavelength. Finding the wavelength,

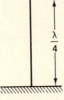

Fig. 6-22

$$f\lambda = kC$$
$$100 \times 10^6 \lambda = 0.90(3 \times 10^8)$$
$$\lambda = \frac{0.90(3 \times 10^8)}{100 \times 10^6}$$
$$= 0.027 \times 10^2$$
$$= 2.7\text{ m}$$

One quarter-wavelength is then

$$\boxed{\frac{\lambda}{4} = \frac{2.7}{4} = 0.675\text{ m}}$$

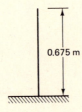

Fig. 6-23

See Fig. 6-23.

6.9 Find the intended frequency of operation of a dipole antenna cut to a length of 3.5 m. Assume a velocity of 3×10^8 m/s for electromagnetic waves on the antenna.

SOLUTION

Given: $L_{\text{dipole}} = 3.5$ m

Find: f_0

A dipole antenna is cut to one half-wavelength. Thus,

$$\lambda = 2L_{\text{dipole}} = 2(3.5) = 7.0\text{ m}$$

Converting the wavelength to frequency,

$$f\lambda = C$$
$$f(7.0) = 3 \times 10^8$$
$$f = \frac{3 \times 10^8}{7}$$
$$= 0.429 \times 10^8$$

$$\boxed{f = 42.9\text{ MHz}}$$

6.10 Determine the dB gain of a receiving antenna which delivers a 40-μV signal to a transmission line over that of an antenna that delivers a 20-μV signal under identical circumstances.

SOLUTION

Given: $V_2 = 40 \ \mu$V

$V_1 = 20 \ \mu$V

Find: A_{dB}

Using the dB-gain formula,

$$A_{dB} = 20 \log_{10} \frac{V_2}{V_1}$$

Substituting,

$$A_{dB} = 20 \log_{10} \frac{40 \times 10^{-6}}{20 \times 10^{-6}}$$

$$= 20 \log_{10} 2$$

Referring to a table of logarithms or using an electronic calculator, $\log_{10} 2 = 0.3$. Thus,

$$A_{dB} = 20(0.3)$$

$$\boxed{A_{dB} = 6 \, \text{dB}}$$

6.11 An antenna that has a gain of 6 dB over a reference antenna is radiating 700 W. How much power must the reference antenna radiate in order to be equally effective in the most preferred direction?

SOLUTION

Given: $A_{dB} = 6$ dB

$P_1 = 700$ W

Find: P_2

The reference antenna must supply an output 6 dB higher than 700 W. Using the dB-gain formula,

$$A_{dB_{10}} = 10 \log_{10} \frac{P_2}{P_1}$$

$$6 = 10 \log_{10} \frac{P_2}{700}$$

$$0.6 = \log_{10} \frac{P_2}{700}$$

$$\text{antilog } 0.6 = \frac{P_2}{700}$$

$$4 = \frac{P_2}{700}$$

$$P_2 = 4(700)$$

$$\boxed{P_2 = 2800 \, \text{W}}$$

6.12   Determine the front-to-back ratio of an antenna which puts out 3 kW in its most optimum direction and 500 W in the opposite direction. See Fig. 6-24.

SOLUTION

Given: $P_F = 3\,\text{kW}$

               $P_B = 500\,\text{W}$

Find: A_{FB}

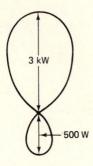

   The front-to-back ratio is the ratio of powers of the optimum direction to its opposite direction expressed in dB.

$$A_{\text{FB}} = 10\log_{10}\frac{P_F}{P_B}$$

$$= 10\log_{10}\frac{3000}{500}$$

$$= 10\log_{10} 6$$

$$= 10(0.778)$$

$$\boxed{A_{\text{FB}} = 7.78\,\text{dB}}$$

Fig. 6-24

6.13   Design a beam antenna consisting of a dipole, one reflector, and one director. Assume a velocity coefficient of 0.90. The antenna is to be cut for a frequency of 150 MHz. See Fig. 6-25.

SOLUTION

Given: $f = 150\,\text{MHz}$

               $k = 0.90$

Find: Beam antenna specifications

   It is first necessary to determine the wavelength of the 150-MHz signal on the antenna.

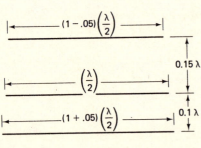

$$f\lambda = kC$$

$$(150 \times 10^6)\lambda = 0.90(3 \times 10^8)$$

$$\lambda = \frac{0.90(3 \times 10^8)}{150 \times 10^6}$$

$$= 1.8\,\text{m}$$

Fig. 6-25

Now, to determine the length of the antenna elements:

$$L_{\text{dipole}} = \frac{\lambda}{2} = \frac{1.8}{2}$$

$$\boxed{L_{\text{dipole}} = 0.9\,\text{m}}$$

$$L_{\text{reflector}} = (1 + 0.05)\frac{\lambda}{2}$$

$$= (1.05)(0.9)$$

$$\boxed{L_{\text{reflector}} = 0.945\,\text{m}}$$

$$L_{\text{director}} = (1 - 0.05)\frac{\lambda}{2}$$

$$= (0.95)(0.9)$$

$$\boxed{L_{\text{director}} = 0.855\,\text{m}}$$

Now determine element spacing:

$$\boxed{0.1\lambda = (0.1)(1.8) = 0.18\,\text{m}}$$

$$\boxed{0.15\lambda = (0.15)(1.8) = 0.27\,\text{m}}$$

See Fig. 6-26.

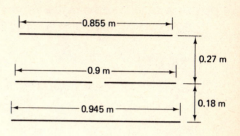

Fig. 6-26

6.14 Determine the optimum frequencies of operation for the antenna shown as Fig. 6-27. Assume a velocity factor of 1.0.

SOLUTION

Given: Antenna shown in Fig. 6-27.

Find: f_1, f_2

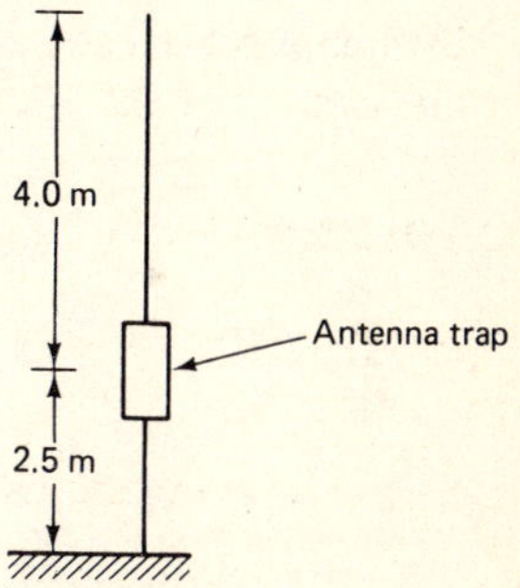

The antenna shown in Fig. 6-27 is a quarter-wave Marconi antenna. The antenna is cut for two frequencies, one whose wavelength is four times the length between the feed point and the antenna trap, and the other four times the total length of the antenna.

$$\lambda_1 = 4 \times 2.5 = 10\,\text{m}$$

$$= 10\,\text{m}$$

$$\lambda_2 = 4(2.5 + 4.0)$$

$$= 26\,\text{m}$$

Fig. 6-27

Converting these wavelengths to frequencies requires the use of the formula

$$f\lambda = kC$$

Thus,

$$f_1\lambda_1 = kC$$

$$f_1(10) = (1)(3 \times 10^8)$$

$$\boxed{f_1 = 30\,\text{MHz}}$$

$$f_2\lambda_2 = kC$$

$$f_2(26) = (1)(3 \times 10^8)$$

$$f_2 = \frac{3 \times 10^8}{26}$$

$$\boxed{f_2 = 11.54\,\text{MHz}}$$

6.15   An antenna is constructed as shown in Fig. 6-28 using antenna traps. Assuming a velocity factor of 1.0, determine the frequencies at which this antenna has been designed to operate.

SOLUTION

Given: Antenna as shown in Fig. 6-28

$k = 1.0$

Find: f_1, f_2

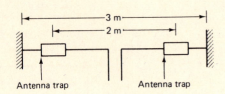

Fig. 6-28

The distance between traps is equal to 1/2 wavelength of the higher optimum frequency of operation. Thus the wavelength of the higher-frequency signal is equal to $2 \times 2\,\text{m} = 4\,\text{m}$.

Converting wavelength to frequency,

$$f_1 \lambda_1 = kC$$
$$f_1(4) = 1(3 \times 10^8)$$
$$f_1 = \frac{3 \times 10^8}{4}$$
$$= 0.75 \times 10^8$$

$$\boxed{f_1 = 75\,\text{MHz}}$$

The full-length distance of the antenna is 1/2 wavelength of the lower optimum frequency of operation. The wavelength of the lower-frequency signal is thus equal to 6 m.

$$f_2 \lambda_2 = kC$$
$$f_2(6) = 1(3 \times 10^8)$$
$$f_2 = \frac{3 \times 10^8}{6}$$
$$= 0.5 \times 10^8$$

$$\boxed{f_2 = 50\,\text{MHz}}$$

6.16   A dipole antenna using antenna traps is desired which is to be used to transmit a 140-MHz signal and a 90-MHz signal. See Fig. 6-29. Design the antenna. Assume a velocity factor of 0.90.

SOLUTION

Given: $f_1 = 140\,\text{MHz}$

$f_2 = 90\,\text{MHz}$

$k = 0.90$

Find: L_1, L_2

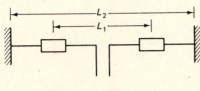

Fig. 6-29

First find the wavelengths of the two signals to be radiated:

$$f_1 \lambda_1 = kC$$
$$140 \times 10^6 \lambda_1 = 0.90(3 \times 10^8)$$
$$\lambda_1 = \frac{0.90(3 \times 10^8)}{140 \times 10^6}$$
$$= 1.929\,\text{m}$$

$$f_2 \lambda_2 = kC$$
$$90 \times 10^6 \lambda_2 = 0.90(3 \times 10^8)$$
$$\lambda_2 = \frac{0.90(3 \times 10^8)}{90 \times 10^6}$$
$$= 3.0 \text{ m}$$

Once the wavelengths of the two signals have been found, the required antenna lengths can be determined:

$$L_1 = \frac{\lambda_1}{2}$$
$$= \frac{1.929}{2}$$

$$\boxed{L_1 = 0.965 \text{ m}}$$

$$L_2 = \frac{\lambda_2}{2}$$
$$= \frac{3.0}{2}$$

$$\boxed{L_2 = 1.5 \text{ m}}$$

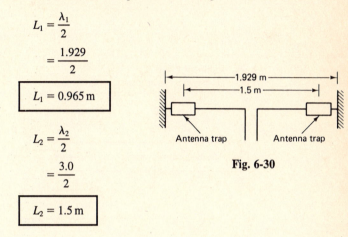

Fig. 6-30

See Fig. 6-30.

6.17 An antenna is to be installed to receive a line-of-sight wave transmitted from an antenna located at a distance of 50 mi from this installation and which is 750 ft in height. Determine the minimum necessary height of the receiving antenna.

SOLUTION

Given: $d = 50$ mi

                $h_t = 750$ ft

Find: h_r

The relationship between antenna height in feet and line-of-sight distance in miles is

$$d = \sqrt{2h_t} + \sqrt{2h_r}$$
$$50 = \sqrt{2(750)} + \sqrt{2h_r}$$
$$50 - 38.73 = \sqrt{2h_r}$$
$$11.27 = \sqrt{2h_r}$$
$$127 = 2h_r$$
$$\frac{127}{2} = h_r$$
$$63.5 = h_r$$

$$\boxed{h_r = 63.5 \text{ ft}}$$

The receiving antenna must be higher than 63.5 ft.

6.18 How far from a transmitting antenna 1500 ft high can a line-of-sight wave be effective? Assume a receiving antenna height of 25 ft. See Fig. 6-31.

SOLUTION

Given: $h_t = 1500$ ft
              $h_r = 25$ ft

Find: d

Using the formula relating distance in miles and antenna height in feet,

$$d = \sqrt{2h_t} + \sqrt{2h_r}$$
$$= \sqrt{2(1500)} + \sqrt{2(25)}$$
$$= \sqrt{3000} + \sqrt{50}$$
$$= 54.77 + 7.07$$

$$\boxed{d = 61.84\,\text{mi, assuming no obstructions}}$$

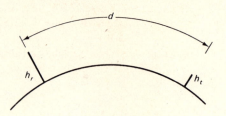

Fig. 6-31

Supplementary Problems

6.19 Sketch the radiation pattern of a Marconi antenna.   *Ans.*   See Fig. 6-1.

6.20 Sketch the radiation pattern of a Hertz antenna.   *Ans.*   See Fig. 6-2.

6.21 What is another name for a Hertz antenna?   *Ans.*   Half-wave dipole.

6.22 Distinguish between the radiation resistance of an antenna and the ohmic resistance of an antenna.   *Ans.*   Radiation resistance accounts for radiated power. Ohmic resistance accounts for heat generated due to resistivity.

6.23 Determine the radiation resistance of an antenna which radiates 1000 W when drawing 5 A.   *Ans.*   40 Ω

6.24 How much current does an antenna draw when radiating 500 W if it has a radiation resistance of 300 Ω?   *Ans.*   1.29 A

6.25 An antenna having a radiation resistance of 50 Ω draws 8 A. How much power is it radiating?
Ans.   3200 W

6.26 Calculate the radiation resistance of an antenna which is fed 12 A when radiating 10 kW.   *Ans.*   69.44 Ω

6.27 How much power does a 50-Ω antenna radiate when fed a current of 4 A?   *Ans.*   800 W

6.28 Determine the Q of an antenna cut for a frequency of 14 MHz having a bandwidth of 4 MHz.   *Ans.* $Q = 3.5$

6.29 What is the bandwidth of an antenna cut for 110 MHz having a Q of 70?   *Ans.*   BW = 1.57 MHz

6.30 An antenna cut for a frequency of 30 MHz has a Q of 40. Determine its bandwidth.   *Ans.*   BW = 750 kHz

6.31 Calculate the length of a half-wave dipole antenna cut for a frequency of 75 MHz. Assume a velocity factor of 0.90.   *Ans.*   1.8 m

6.32 Assuming a velocity factor of 0.88, determine the optimum length of a Marconi antenna which is to be cut for a frequency of 65 MHz.   *Ans.*   1.015 m

6.33 Calculate the required length of a half-wave dipole antenna cut for 90 MHz if the velocity factor on the antenna is 0.87. Then determine the required length of the antenna if the velocity factor were 1.0.
Ans.   1.45 m, 1.667 m

6.34 A Marconi antenna is 2.8 m in length. The velocity factor is 0.80. What frequency was this antenna cut for?   *Ans.*   21.43 MHz

6.35 An experimental antenna has a gain of 3 dB above a reference antenna. How much power would the reference antenna have to radiate in order to provide the same signal picked up when 5 W is radiated by the experimental antenna?   *Ans.*   10 W

6.36 The same test signal is to be radiated by antenna A and then by antenna B. A receiving antenna picks up a 5-μV/m signal when the transmission is broadcast by antenna A. When the same test signal is broadcast by antenna B, the receiving antenna picks up a 20-μV/m signal. Calculate the gain of antenna B over antenna A.   *Ans.*   12 dB

6.37 An antenna having a gain of 3 dB over a reference antenna is radiating 1500 W. How much power must the reference antenna radiate in order to be equally as effective in the most preferred direction?   *Ans.*   3000 W

6.38 Calculate the front-to-back ratio of an antenna which radiates 500 W in a northerly direction and 50 W in a southerly direction.   *Ans.*   10 dB

6.39 Sketch an antenna showing one driven element, one reflector, and one director. Assume the speed of propagation of the electromagnetic wave on the antenna to be 3×10^8 m/s. Determine the length of each of the antenna elements and the distance between them if this antenna is optimized for 40 MHz.
Ans.   3.75 m, 3.9375 m, 3.5625 m, 0.75 m, 1.125 m

6.40 Design a three-element beam antenna (one driven element, two parasitic elements) for use with a 200-MHz signal. Consider that the velocity factor on the antenna is 0.85.
Ans.   0.6375 m, 0.669 m, 0.606 m, 0.1275 m, 0.1913 m

6.41 The driven element of a three-element Yagi antenna is to be 3.5 m in length. The velocity factor for this antenna is 0.89. Determine the frequency for which this driven element is cut and calculate the length and spacing of the other antenna elements.   *Ans.*   38.14 MHz, 3.675 m, 3.325 m, 0.7 m, 1.05 m

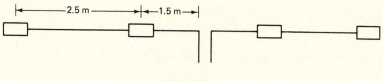

Fig. 6-32

6.42 An antenna is constructed as shown in Fig. 6-32. At what frequencies is this antenna designed to operate? Assume a velocity factor of 1.0.   *Ans.*   50 MHz, 18.75 MHz

6.43 Design a dipole using traps for transmitting signals at 175 MHz and 100 MHz and trap locations. Assume a velocity factor of 1.0.   *Ans.*   0.429 m, 0.321 m. Show segment lengths

6.44   Determine the optimum frequencies of operation for the Marconi antenna shown in Fig. 6-33. Assume a velocity factor of 1.0.   *Ans.*   25 MHz, 9.375 MHz

6.45   Determine the maximum effective distance of the line-of-sight wave of a transmission being radiated from an antenna whose height is 1200 ft if the receiving antenna height is 100 ft.   *Ans.*   63.13 mi

6.46   Calculate the necessary height of an antenna to receive a line-of-sight wave being transmitted from an antenna 1000 ft in height which is located 100 mi from the receiving antenna. *Ans.*   1527.9 ft

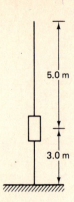

5.0 m

3.0 m

Fig. 6-33

Index